AF789992

Sustainable Energy Investment - Technical, Market and Policy Innovations to Address Risk

Edited by Joseph Nyangon and John Byrne

Published in London, United Kingdom

IntechOpen

Supporting open minds since 2005

Sustainable Energy Investment - Technical, Market and Policy Innovations to Address Risk
http://dx.doi.org/10.5772/intechopen.86093
Edited by Joseph Nyangon and John Byrne

Contributors
Tam Kemabonta, Abd-alla Gad, Olivia Muza, Marco Tortora, Giuseppe Tortora, Dulce Esmeralda Garcia Ruiz, Jorge Alberto Navarro Serrano, Job Taminiau, John Byrne, Daniel Sanchez Carretero, Soojin Shin, Jing Xu, Cecilia Camporeale, Mario Jorizzo, Roberto Delciello, Ryan Merlin Yonk, Anneliese Gegenheimer, C. Michael Gegenheimer, Joseph Nyangon, Tahira Shafique, Javeria Shafique

© The Editor(s) and the Author(s) 2021
The rights of the editor(s) and the author(s) have been asserted in accordance with the Copyright, Designs and Patents Act 1988. All rights to the book as a whole are reserved by INTECHOPEN LIMITED. The book as a whole (compilation) cannot be reproduced, distributed or used for commercial or non-commercial purposes without INTECHOPEN LIMITED's written permission. Enquiries concerning the use of the book should be directed to INTECHOPEN LIMITED rights and permissions department (permissions@intechopen.com).
Violations are liable to prosecution under the governing Copyright Law.

(cc) BY-NC

Individual chapters of this publication are distributed under the terms of the Creative Commons Attribution - NonCommercial 4.0 International which permits use, distribution and reproduction of the individual chapters for non-commercial purposes, provided the original author(s) and source publication are appropriately acknowledged. More details and guidelines concerning content reuse and adaptation can be found at http://www.intechopen.com/copyright-policy.html.

Notice
Statements and opinions expressed in the chapters are these of the individual contributors and not necessarily those of the editors or publisher. No responsibility is accepted for the accuracy of information contained in the published chapters. The publisher assumes no responsibility for any damage or injury to persons or property arising out of the use of any materials, instructions, methods or ideas contained in the book.

First published in London, United Kingdom, 2021 by IntechOpen
IntechOpen is the global imprint of INTECHOPEN LIMITED, registered in England and Wales, registration number: 11086078, 5 Princes Gate Court, London, SW7 2QJ, United Kingdom
Printed in Croatia

British Library Cataloguing-in-Publication Data
A catalogue record for this book is available from the British Library

Additional hard and PDF copies can be obtained from orders@intechopen.com

Sustainable Energy Investment - Technical, Market and Policy Innovations to Address Risk
Edited by Joseph Nyangon and John Byrne
p. cm.
Print ISBN 978-1-83880-197-7
Online ISBN 978-1-83880-198-4
eBook (PDF) ISBN 978-1-83962-508-4

An electronic version of this book is freely available, thanks to the support of libraries working with Knowledge Unlatched. KU is a collaborative initiative designed to make high quality books Open Access for the public good. More information about the initiative and links to the Open Access version can be found at www.knowledgeunlatched.org

For EU product safety concerns:
IN TECH d.o.o., Prolaz Marije Krucifikse Kozulić 3, 51000 Rijeka, Croatia, info@intechopen.com or visit our website at intechopen.com.

We are IntechOpen, the world's leading publisher of Open Access books

Built by scientists, for scientists

5,200+
Open access books available

128,000+
International authors and editors

150M+
Downloads

156
Countries delivered to

Our authors are among the
Top 1%
most cited scientists

12.2%
Contributors from top 500 universities

WEB OF SCIENCE™

Selection of our books indexed in the Book Citation Index in Web of Science™ Core Collection (BKCI)

Interested in publishing with us?
Contact book.department@intechopen.com

Numbers displayed above are based on latest data collected.
For more information visit www.intechopen.com

Meet the editors

Dr. Joseph Nyangon is a senior researcher at the Center for Energy and Environmental Policy, University of Delaware. He is also a senior research fellow at the Foundation for Renewable Energy and Environment, a non-resident fellow of the Payne Institute at the Colorado School of Mines, and a research fellow in the Initiative for Sustainable Energy Policy at the Johns Hopkins University's School of Advanced International Studies (SAIS). He holds a Ph.D., two master's degrees, and an undergraduate degree focusing on energy economics, public policy, energy systems engineering and computing systems from Columbia University, the University of Delaware, among others. Dr. Nyangon's practice focuses on applying optimization methods and econometric modeling techniques to evaluate electricity systems and generate insights to inform policy, risk pricing strategies, and planning decisions.

Dr. John Byrne has contributed to Working Group III of the United Nations-sponsored Intergovernmental Panel on Climate Change (IPCC) since 1992. Dr. Byrne is an advisor to the "Solar City Seoul" initiative, which is building 1 GWp of solar power on public buildings, parking facilities, and residential and commercial buildings. He is co-editor-in-chief of the invitation-only journal, *WIREs Energy and Environment*. He has published nineteen books and more than 170 research articles.

Contents

Preface

This book presents, frames, and addresses both renewable energy investment risks and asset risks associated with conventional energy technology. In particular, it examines the technical, market, and policy strategies needed to mitigate risks of energy investments against climate change impacts and how governance systems can be structured to deal with stranded asset risks resulting from repricing or write-downs of carbon-intensive assets. Helping governments and organizations to address various risk dimensions, improve energy efficiency, and promote sustainable energy investments in project development is fully aligned with the goals of the 2015 Paris Agreement and the central mandate and purpose of the United Nations' Sustainable Development Goals (SDGs). The book is structured as follows:

- Part I is the Introduction
- Part II deals with risk dynamics, principles, and drivers characterizing renewable energy investments.
- Part III details emerging energy innovation agendas being developed and implemented globally to support sustainable energy technologies.
- Part IV analyzes a broad range of case studies from developing countries—Rwanda, Algeria, Egypt, Pakistan, and Mexico—that could accelerate sustainable energy investment and advance inclusive energy transformation in line with global climate and sustainable development objectives.

This book offers a comprehensive reference for those interested in climate-sensitive sustainable energy investment. Its content should appeal to companies, students, researchers, academics, and policymakers interested in long-term trends in climate and stranded-asset risks, risk management and mitigation, climate-proofing energy infrastructure, energy markets and innovation, and energy policy development.

Joseph Nyangon and John Byrne
Center for Energy and Environmental Policy,
University of Delaware,
USA

Acknowledgements

We are grateful for this chance to contribute to the energy sector's transition, and to the colleagues at the Center for Energy and Environmental Policy for stimulating conversations and scientific, economic, and policy insights. Finally, we would like to thank Ana Pantar, Edi Lipovic, and Sandra Maljavac of IntechOpen who made this publication possible.

Section 1

Introduction

Chapter 1

Introductory Chapter: Sustainable Energy Investment and the Transition to Renewable Energy-Powered Futures

Joseph Nyangon and John Byrne

1. Introduction

"Sustainable energy investment" is a widely used phrase and concept in the fields of finance, engineering and economics. Typically, it focuses on evaluating renewable power development and includes assessments of political and regulatory risks, energy risk hedging and portfolio diversification. Often publications on this topic contribute to the climate change response agenda: promote investments in solar- or wind-powered technologies in order to realize a more equitable, sustainable and prosperous future; evaluate financial aspects of carbon budgeting and energy asset risk management; and respond to financial and climate risks associated with mitigation and adaptation policy interventions. Policymakers and energy regulators correctly perceive climate change to pose threats to energy assets, research and development (R&D), technological innovation to accelerate energy transitions and these impacts are projected to grow in the coming decades [1–3]. Concurrently, the energy sector is experiencing a myriad of challenges, from aging infrastructure, retiring workforces, years of stagnant investment to the need to attract new investment in smart grid resilience, business model innovation reforms, changing customer expectations, and more recently COVID-19 forced disruptions [4, 5]. To mitigate the worst possible impacts, attention is now shifting to strategies for de-risking energy investments—for example, long-term climate-risk hedging and adaption strategies in energy infrastructure development around financing, costs, and revenue—to foster local, national and supranational systems of resource autonomy and reduce the risks of climate change [6–9].

Mainstreaming renewable-powered energy investment into business decision-making and risk pricing is an attractive climate-smart solution that societies and economies can adopt immediately to help overcome anachronistic electric power regimes and regional development dynamics. Globally, investments in distributed, renewable energy-powered futures keep accelerating with clear upward trends in worldwide power generation expansion and risk management at the forefront. Nevertheless, finding a low-carbon, risk pricing formula is not easy. Despite compelling arguments for investment in low-carbon technologies and applications, such as small-scale renewables and locally distributed green energy, digitalization, advanced batteries or carbon capture and storage (CCS), these interventions require a pragmatic assessment of their financeability, which in turn hinges on their technical and economic potential with respect to complex factors, including social equity, feasibility, socioeconomic impact, and climate impact. The scale of

IntechOpen

deploying these low-carbon technologies is also an important consideration to investors because project size is a critical determinant of the cost of unit returns. Some institutional investors consider small-scale investments less attractive due to their perceived low rate of returns. On the other hand, large-scale power systems, such as grid-scale battery storage and other scalable carbon-free power technologies require significant investment in risk hedging and portfolio diversification [8, 10]. The two principal risks that are often mentioned in this area involve (i) those arising from the physical effects of climate change on energy infrastructure, institutions, business operations, energy markets and assets, and (ii) risks resulting from investment in zero-carbon transition strategies due to changes in technology, policy, legal, and market factors. **Table 1** summarizes various dimensions of these renewable energy investment risks.

Suitable climate-smart development—combining innovation mix in technology with those in policy development, new business models, systems operations and market design innovation—could do much to keep the global temperature within the 2 °C carbon budget [3, 23]. Mature non-hydro power sources of renewable

Dimension	Risk factor	References
Technological risk	• R&D capacity • Technology maturity, innovation and progressiveness • Alternative technology	[2, 13–15]
Political risk	• Political stability (internal and external conflicts) • Land acquisition risk • Government credit or foreign debts • Bribery and corruption indices • Legislative and administrative actions • Property rights • Transparency and accountability	[9, 16–19]
Economic foundation and market risk	• Gross domestic product per capita • Exchange rate stability • National/regional economic development level • Contract change risk • Market fluctuations • Change in taxes	[2, 16–18, 20]
Resource risk	• Solar PV and solar thermal potential • Hydropower potential • Wind power potential • Biomass power potential • Geothermal power potential	[8, 13, 20–22]
Environmental / social risk	• Cultural difference • Social cohesion, instability and public resistance • Influence on local environment • Energy demand • Force majeure	[3, 7, 17, 18, 20]

Table 1.
Dimensions of renewable energy investment risk management [11, 12].

electricity, such as solar photovoltaics and onshore wind that can be deployed in a wide range of operating conditions, are generally considered low-risk. These technologies attract large-scale investments and deployment globally, but they are sometimes situated in challenging geographical locations and are vulnerable to weather conditions changes. For example, the risk of technical failure due to extreme weather conditions is always present. Risk averse institutional investors prefer investing in energy technologies with higher rate of return, improved reliability, and operational.

On the other hand, early-stage crucial technologies that have the potential to provide step-change reductions in both cost and energy requirements, and are not as vulnerable to weather and other external events. For example, CCS, and offshore wind are characterized by several technical and financial uncertainties, and are considered high-risk investments by some investors. Typically, investment in such new technologies is often characterized by a 'wait and see' approach to allow them to undergo deployment cycles before they can attract long-term investment commitments.

2. Reimagining sustainable energy investment

Expectations about the market, policy and technological impact of sustainable energy investment continue to evolve. As evidence of the impact of climate change intensifies, consumers and communities are taking action to support a clean energy future and address institutional capacity gaps in public and private investments that hinder the value of decarbonization and adoption of smart energy frameworks [22]. To this end, governments and organizations are beginning to prioritize key innovations that promote clean energy investment readiness. Some of these actions include:

- Planning for energy infrastructure investment, technology transfer and R and D consistent with net-zero transition goal.
- Repurposing existing fossil fuel infrastructure to reduce the overall cost of transition by applying machine learning and artificial intelligence-supported technologies to tackle the risks of stranded electricity assets.
- Delivering finance for electrification of buildings and building automated performance control to raise energy savings performance guarantee and mitigate energy efficiency risks and a possible rebound effect [14].
- Catalyzing finance and investment flows in technology transfer and development in different configurations within and between institutions and across national, sub-national and local levels.
- Supporting communities, businesses and workers by promoting fair access to long-term innovative financing mechanisms and employment opportunities.
- Applying analytics to improve monitoring, reporting, and verification of clean energy technology transfer and investment in R&D.
- Retooling clean energy investment readiness and loan guarantee programs to improve investment flows, project investability and market efficiency.

Reversing the steady growth of greenhouse gas emissions to reduce the deleterious risk of climate change remains the biggest challenge of modern times [1, 2]. It requires a complete change in basic assumptions on how we produce, deliver and consume energy, for example, by focusing on low- or no-carbon energy technologies and greater energy efficiency deployment instead of heavy reliance on energy system dominated by fossil fuel combustion [14]. This vision conflicts with existing socioeconomic growth paradigm. Typically, there exists strong correlation between energy consumption and economic development—especially among developing nations, where poverty reduction strategies are often modeled against increased economic growth, which results in greater energy demand. This relationship is captured by IPAT (I = PAT) model which emphasizes three main factors affecting the environment, i.e., the environmental impact (I) and its relation to population (P), affluence (A), and technology (T) [24].

The emerging shift toward net-zero carbon dioxide (CO_2) emissions development presents an alternative economic development paradigm that could break this strong linkage between economic growth and environmental pollution or CO_2 emissions. For example, to realize high penetration of renewable electricity generation, a clear long-term pathway exists in reimagining the electric grid. Three pillars of this reimagined grid include (a) decarbonizing the electric power supply through the growth of carbon-free power generation sources to improve reliability, affordability and environmental impact of the electricity, and stimulate local economic development, (b) electrification of transportation and buildings, and (c) sequestering the remaining carbon through carbon capture technologies [8, 12, 14–16, 22]. These are vital socioeconomic goals achieved through investment in inverter-based resources (solar, wind, and energy storage).

This book discusses these core dimensions of renewable energy-powered innovations and investment risk. It is a selective compilation of climate-sensitive working concepts, technological solutions and country-specific case studies positioned within the broader debate of just energy transitions. The volume contributes to the existing body of knowledge needed to accelerate renewable energy deployment to meet rising energy demand and ensure that the transition is global, inclusive, socially equitable, and more sustainable.

Author details

Joseph Nyangon* and John Byrne
University of Delaware, USA

*Address all correspondence to: jnyangon@gmail.com

IntechOpen

© 2021 The Author(s). Licensee IntechOpen. Distributed under the terms of the Creative Commons Attribution - NonCommercial 4.0 License (https://creativecommons.org/licenses/by-nc/4.0/), which permits use, distribution and reproduction for non-commercial purposes, provided the original is properly cited. CC BY-NC

References

[1] Byrne J, Hughes K, Rickerson W, Kurdgelashvili L. American policy conflict in the greenhouse: Divergent trends in federal, regional, state, and local green energy and climate change policy. Energy Policy. 2007; **35**(9): 4555-4573

[2] Curtin J, McInerney C, Ó Gallachóir B, Hickey C, Deane P, Deeney P. Quantifying stranding risk for fossil fuel assets and implications for renewable energy investment: A review of the literature. Renewable and Sustainable Energy Reviews. 2019; **116**: 109402

[3] Nyangon J. Tackling the risk of stranded electricity assets with machine learning and artificial intelligence. In: Nyangon J, Byrne J, editors. *Sustainable Energy Investment: Technical, Market and Policy Innovations to Address Risk*. London, UK: Intech Open; 2020. pp. 1. 22

[4] Costello KW, Hemphill RC. Electric utilities' 'death spiral': Hyperbole or reality? Electricity Journal. 2014;**27**(10):7-26

[5] Fox-Penner PS. *Smart Power: Climate Change, the Smart Grid, and the Future of Electric Utilities*. 1st ed. Washington: Island Press; 2010 ISBN: 9781597267052.

[6] Byrne J, Taminiau J. Utilizing the urban fabric as the solar power Plant of the Future. In: Droege P, editor. *Urban Energy Transition*. 2nd ed. New York City: Elsevier; 2018. pp. 31-49. DOI: https://doi.org/10.1016/B978-0-08-102074-6.00016-4

[7] Clapp C, Sillmann J. Facilitating climate-smart investments. One Earth. 2019;**1**(1):57-61

[8] Fox-Penner P. *Power after Carbon: Building a Clean, Resilient Grid*. Cambridge, MA: Harvard University Press; 2020 ISBN: 978-0674241077

[9] Yang K, Wei Y, Li S, He J. Geopolitical risk and renewable energy stock markets: An insight from multiscale dynamic risk spillover. Journal of Cleaner Production. 2021;**279**:123429

[10] Capellán-Pérez I, de Castro C, Miguel González LJ. Dynamic energy return on energy investment (EROI) and material requirements in scenarios of global transition to renewable energies. Energy Strategy Reviews. 2019;**26**:100399

[11] Gatzert N, Vogl N. Evaluating investments in renewable energy under policy risks. Energy Policy. 2016;**95**:238-252

[12] van der Ploeg F, Rezai A. Stranded assets in the transition to a carbon-free economy. Annual Review of Resource Economics. 2020;**12**(1)

[13] Liu X, Zeng M. Renewable energy investment risk evaluation model based on system dynamics. Renewable and Sustainable Energy Reviews. 2017;**73**:782-788

[14] Nyangon J, Byrne J. Spatial energy efficiency patterns in New York and implications for energy demand and the rebound effect. Energy Sources, Part B: Economics, Planning, and Policy. 2021. https://doi.org/10.1080/15567249.2020.1868619

[15] Rubin ES, Azevedo IML, Jaramillo P, Yeh S. A review of learning rates for electricity supply technologies. Energy Policy. 2015;**86**:198-218

[16] Nyangon J. Distributed Energy Generation Systems Based on Renewable Energy and Natural Gas Blending: New Business Models for Economic Incentives, Electricity Market Design and Regulatory Innovation. Ph.D. Dissertation, College of Engineering, University of Delaware. Delaware: Newark; 2017

[17] Nyangon J, Alabbas N, Agbemabiese L. Entangled Systems at the Energy-Water-Food Nexus: Challenges and opportunities. In: Rao P, Patil Y, editors. *Reconsidering the Impact of Climate Change on Global Water Supply, Use, and Management*. United States: IGI Global; 2017a. pp. 145-165

[18] Nyangon J, Byrne J, Taminiau J. Assessment of Price convergence between natural gas and solar photovoltaic in the U.S. electricity market. *Wiley Interdisciplinary Reviews: Energy and*. Environment. 2017b;**6**(3):1-20

[19] Shimbar A, Ebrahimi SB. Political risk and valuation of renewable energy investments in developing countries. Renewable Energy. 2020;**145**:1325-1333

[20] Tietjen O, Pahle M, Fuss S. Investment risks in power generation: A comparison of fossil fuel and renewable energy dominated markets. Energy Economics. 2016;**58**:174-185

[21] Duan F, Ji Q, Liu B, Fan Y. Energy investment risk assessment for nations along China's Belt & Road Initiative. Journal of Cleaner Production. 2018;**170**:535-547

[22] Nyangon J. Smart energy frameworks for smart cities: The need for polycentrism. In: Augusto JC, editor. *Handbook of smart cities* (pp. 1-32). Cham: Springer. ISBN: 978-3-030-15145-4. DOI. 2021. https://doi.org/10.1007/978-3-030-15145-4_4-2

[23] McGlade C, Ekins P. The geographical distribution of fossil fuels unused when limiting global warming to 2 °C. Nature. 2015;**517**(7533):187-190

[24] Koçak E, Ulucak ZŞ. The effect of energy R&D expenditures on CO2 emission reduction: estimation of the STIRPAT model for OECD countries. Environmental Science and Pollution Research. 2019;26(14):14328-14338.

Section 2

Risk Dynamics, Principles and Drivers

Chapter 2

Tackling the Risk of Stranded Electricity Assets with Machine Learning and Artificial Intelligence

Joseph Nyangon

Abstract

The Paris Agreement on climate change requires nations to keep the global temperature within the 2°C carbon budget. Achieving this temperature target means stranding more than 80% of all proven fossil energy reserves as well as resulting in investments in such resources becoming stranded assets. At the implementation level, governments are experiencing technical, economic, and legal challenges in transitioning their economies to meet the 2°C temperature commitment through the nationally determined contributions (NDCs), let alone striving for the 1.5°C carbon budget, which translates into greenhouse gas emissions (GHG) gap. This chapter focuses on tackling the risks of stranded electricity assets using machine learning and artificial intelligence technologies. Stranded assets are not new in the energy sector; the physical impacts of climate change and the transition to a low-carbon economy have generally rendered redundant or obsolete electricity generation and storage assets. Low-carbon electricity systems, which come in variable and controllable forms, are essential to mitigating climate change. These systems present distinct opportunities for machine learning and artificial intelligence-powered techniques. This chapter considers the background to these issues. It discusses the asset stranding discourse and its implications to the energy sector and related infrastructure. The chapter concludes by outlining an interdisciplinary research agenda for mitigating the risks of stranded assets in electricity investments.

Keywords: stranded assets, stranded resources, unburnable carbon, machine learning, artificial intelligence, carbon budgets, derisking investments, climate change

1. Introduction

The power industry is in transition, and energy management systems are adapting to it. Recently, the rapid proliferation of distributed energy resources (DERs) (e.g., distributed generation such as residential solar photovoltaics (PV) and wind electricity, controllable loads, and energy storage), have transformed operational, planning, and regulatory dynamics. Low-cost natural gas in the US, Europe, and elsewhere continues to push gas-fired electricity generation to the top of the generation mix. To this end, governments continue to promote low-carbon technologies through ever-stringent energy policies, like renewable portfolio standards (RPS), net metering, feed-in tariffs, and carbon pricing initiatives and emission trading schemes like the European Union Emission Trading System (EU ETS), Switzerland Emissions Trading Scheme, emissions trading schemes in China and Australia, the Regional Greenhouse Gas Initiative (RGGI) in the nine U.S. states in the Northeast and Mid-Atlantic region, the Transportation and Climate Initiative (TCI) under consideration for transportation emissions in the Northeast and Mid-Atlantic, the California and Quebec's Western Climate Initiative, among others. Furthermore, this growth in renewable electricity generation has been motivated by customers' preference for distributed energy as a means to fostering grid reliability and system efficiency, cost reduction, and improved customer choice over their power supplies [1–3].

These efforts are in line with the 2015 Paris Agreement on climate change and its nationally determined contributions' (NDCs) long-term goal of keeping the rise in global mean temperature to "well below [two degrees Celsius (2°C)] above pre-industrial levels and pursuing efforts to limit the temperature increase to 1.5°C above pre-industrial levels" [4]. Moreover, limiting these temperature targets requires reaching net-zero global carbon dioxide (CO_2) emissions "between 2060 and 2070," with full decarbonization or "net negative CO_2 emissions" realized before the end of the century [5]. From a policy perspective, this transition from carbon-intensive sources to low and non-carbon-emitting sources is continuing as high penetrations of distributed electricity, energy storage and management devices, and investment in new forms of flexible demand resources become connected to the power grid network. Such significant shifts threaten the fossil energy business model and could, in turn, result in the "stranding" of the carbon-intensive assets through retirement or devaluation [6–8]. In other words, meeting the Paris temperature targets necessitates turning existing fossil fuel investments into stranded assets and fossil fuel reserves into stranded resources. The concept of "stranding" or "stranded assets" has been explored broadly in extant literature, from environment-related risk exposure of coal assets [9] to "unburnable fossil fuel deposits" such as oil, gas, and coal and the risk of stranded assets [10, 11].

Bos and Gupta [12] define stranded assets as "assets that lose economic value well ahead of their anticipated useful life, whether that is a result of changes in legislation, market forces, disruptive innovation, societal norms, or environmental shocks" p. 1 and stranded resources as "resources which are considered uneconomic or cannot be developed or extracted as a result of technological, spatial, regulatory, political or market limitations, or changes in social and environmental norms" p. 2. On the other hand, Caldecott et al. [13] define stranded assets as those assets which "suffer from unanticipated or premature write-offs, downward revaluations or [conversion] to liabilities" p. 11. Policymakers and experts concur that this transition should be managed proactively and pragmatically because if done haphazardly, it could perpetuate the techno-institutional complex of "carbon lock-in" and path dependency, thus making future transitions difficult [14–21]. On the other hand, if variable renewable energy resources like solar and wind electricity is introduced in significant quantities and not correlated exactly with peak load, it may create a unique challenge like the infamous California ISO's "duck curve" shown in **Figure 1**. How should the energy sector respond?

This chapter is structured as follows. Section 2 discusses digitalization solutions and business model innovations and presents moral arguments for supply- and demand-side energy solutions, including sensors, meters, higher efficiency devices, and energy auditing, including measurement and verification strategies that can be utilized to improve energy management. Specifically, using ML (machine learning- and (AI) artificial intelligence solutions to support (a) tackling stranded assets in

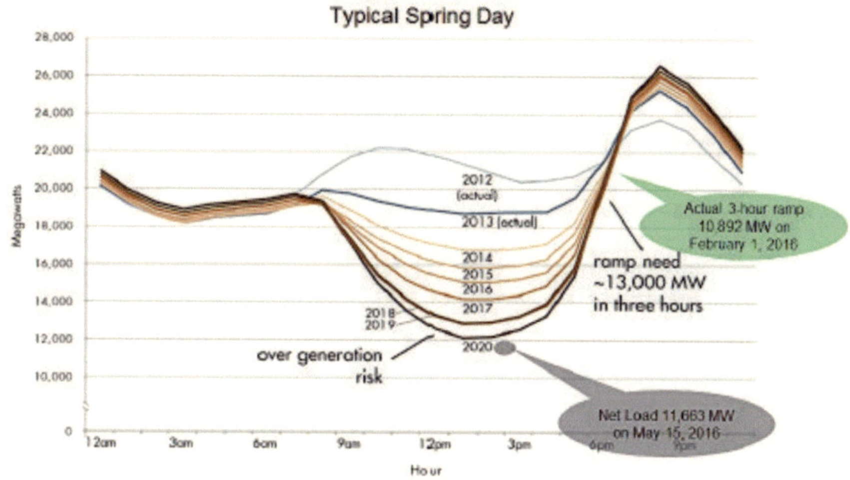

Figure 1.
California ISO's "duck curve." Source: CAISO (2014).

the transition to a low-carbon economy (b) real-time measurement of energy data, (c) manage data gathering and monitoring, (d) and proactively and accurately analyze the data gathered to detect changes in supply-demand imbalances and improve the situation promptly. Section 3 reviews the risks of stranding and identifies distinct opportunities for ML and AI applications in the energy sector. Section 4 emphasizes the impact of stranding risk factors on oil, gas, and coal resources and how this translates into the concept of "unburnable fossil fuel deposits." Section 5 discusses how advances in ML and AI techniques might help tackle the risks of stranded carbon assets, and Section 6 concludes.

2. Leveraging digitalization and business model innovations for energy management

Today's modern cities are sprouting with new industrial buildings and residential complexes. The consensus is emerging that dramatic growth in distributed renewable energy, and digitalization in economy and innovations, two megatrends of the twenty-first century, are critical strategies for climate change mitigation and changing the greenhouse gas (GHG) emission trajectories. Yet, while the electric power system is in transition, many of the vital power system challenges which confront governments and businesses, like access to a cleaner, more resilient, reliable, and affordable electricity, remain underfunded and unresolved. The increased deployments of energy management applications across the transportation, buildings, and industrial sectors, for example, reduce the cost of operation and consumption, lower energy losses, increase grid reliability, improve electric power production from carbon-free sources, and alleviate investment inefficiencies that could cause an energy-efficiency gap [22–24]. The continued growth of the fluctuated distributed generations (such as solar photovoltaics, wind turbines, electric vehicles, and energy storage systems) may perturb the network and create voltage drop/rise problems and in severe conditions, blackouts.

In a highly electrified economy with high shares of variable solar and wind electricity systems, reducing systemic mismatches between the generation and energy demand assets in an efficient manner requires investments in smart energy management systems. Energy management systems consist of two main categories: (a) supply-side devices from the electric utility-side used to manage the fluctuation of the load demand such as substations, and (b) the demand-side management devices used to manage energy consumption and meet the available power from the generation side [25–28]. Substations encompass transformers, switchgear, and protection, control and automation systems, and connect parts of the electric grid that operate at different voltage levels and managing these multidirectional power flows while ensuring reliability and security is critical as the share of decentralized and renewable energy increases. The rise of smart energy management systems, including ML, AI, big data, smart sensors, and the Internet of Things, is a boon not only to the electric power industry—especially in reducing operational costs and carbon emissions—but also the energy transition. For example, opportunities for leveraging digitalization for business model innovation in smart energy management and the corresponding implications for the power sector are substantial and untapped [29, 30].

Energy management is subject to barriers and limitations, which can delay its full market integration. These barriers include high cost of system implementation, inflexible fixed-price electricity tariff system and rate design, aging network's infrastructure, and the need for bidirectional power flow, which is ideal for an intelligent grid network. As a result, energy management continues to have a prominent role in decarbonization. Using ML algorithms and AI optimization models, utilities and system operators can apply optimal dynamic pricing and energy storage resource to improve the management of the "duck curve" phenomenon. For example, Sheha et al. [31] applied game-theoretic models to show that leveraging a combined effect of dynamic pricing profiles and distributed electrical energy storage can help flatten the duck curve, thus solar energy can be increasingly added on the grid without resulting in grid failure. The duck curve problem arises when increasing solar penetration on the grid creates a dip in net load in the middle of the day as solar generation peaks and wind electricity is low, followed by a significant rise in peak in residential demand at sunset as, without any form of energy storage, solar electricity rapidly subsides, and customer consumption increases as citizens get home from work/school thus necessitating significant ramping of thermal generators [24]. **Figure 1** shows California Independent System Operator's (CAISO) widely known "duck curve" (**Figure 1**) [32]. Besides California, the "duck curve" phenomenon also occurs in energy markets with high solar electricity penetrations such as Italy, Germany, Hawaii, and others.

To eliminate the risk of over-generation and possibly, alleviate the "duck curve" problem, implementation of long-term solutions focusing on distinct opportunities for ML techniques, including distributed solar coupled with storage technologies and smart energy management, are emerging in various energy markets. At stake, according to Guidehouse Insights (formerly Navigant Research), is $278 billion in annual global market for the deployment of commercial and industrial (C&I) energy as a service (EaaS) solutions by 2028 [33].

3. The risks of stranding in the energy sector

The intergenerational issues associated with climate change identifies it as an externality associated with carbon dioxide and other GHG emissions because it involves costs that are borne by future generations who have not created the emissions [34–38]. Climate change economists have introduced the concept of "social costs of carbon," which externalizes the externalities of these emissions by denoting the damages caused by them with a monetary value [35, 39–41]. It is for this reason that climate policy experts have advocated for a carbon price to achieve the "right price" as well as incentivize the investments in low-carbon technologies. Furthermore, from a policy, equity and regulatory point of view, scaling the deployment of low-carbon energy technologies inspires innovation in technological development, diffusion, transfer, and discourages the holding of dirty exhaustible assets (fossil fuel reserves) [42, 43], which are prone to becoming stranded due to perfect substitution, and disproportionately impact low- and moderate-income communities.

The risks of stranding of assets are likely to occur during the transition to a green economy. As van der Ploeg and Rezai [17] suggest two conditions are necessary for this transition to occur: (1), the unexpected future changes in the conditions likely to affect the economics of fossil fuel assets, such as customer demand, the social cost of carbon that values the climate externalities, and equity and efficiency considerations, must be present; (2) the cost of shifting around "the underlying capital stocks in the carbon-intensive industries to productive use elsewhere after the energy transition" must be too prohibitive or impossible to meet. Expectations about stranding carbon-intensive assets can occur due to sudden policy change, a breakthrough innovation in renewable energy technology such as energy storage batteries, which can lead to the stranding of fossil fuel-based financial assets since they directly pose a threat to the sustainability of the coal, oil, and gas-based business model.

With ML and AI techniques, energy operators can foster better short-term and long-term forecasting to improve electricity scheduling and integrated system planning, respectively. This would enable the utility operators and system managers to reduce their reliance on polluting, exhaustible fossil assets as well as proactively manage increasing amounts of distributed, low-carbon, variable energy sources like solar and wind energy. Additionally, the ML-AI-driven energy forecasts can provide accurate and optimal management of power grid fluxes to help operators proactively match demand-supply imbalances, manage uncertainties, as well as understand where, when and how many solar power systems [44, 45] and wind generation plants [46] should be built.

However, much of these forecasts employ domain-agnostic techniques, in which domain-specific scenarios are often less applied. For this reason, ML and AI algorithms of the future must incorporate weather-related innovations in climate science and weather modeling techniques in order to improve parametric and nonparametric estimates of both short- and long-term forecast uncertainty, for example, of variable generation and electricity demand [44, 46–49]. For example, using a novel deep learning framework that combines wavelet transforms, stacked autoencoders, and long-short term memory, Bao et al. [50] produced stock price forecasts that outperform other similar models in both predictive accuracy and profitability performance. This notion can be extended to aid electricity demand forecasts that optimize intraday and day-ahead levelized cost and levelized avoided cost of electricity generation resources that minimize GHG emissions. More broadly, in the transition from the incumbent centralized electricity network to a distributed model that is underway, driven by the rapid growth of DERs, understanding the domain value of improved forecasts (e.g., to model electricity load in rural microgrids) across the quartiles of electricity market operation, matching of supply-demand imbalances, network control, and governance and administrative networks [2, 51] is an exciting challenge for ML and the debate on stranded assets.

4. Investments, stranding risk factors, and unburnable fossil fuel deposits

Going by Bos and Gupta [12] and Caldecott et al. [9]'s definition, stranded assets and stranded resources manifest in two main ways (1) devaluation through the unburnable fossil fuel resources, which must be kept in the ground to keep the long-term global temperature target to below 2°C above pre-industrial levels [4], and (2) premature retirement of exhaustible fossil capital assets due to climate policies, including the optimal social cost of carbon in the form of a carbon price [40]. **Figure 2** shows the US annual electricity generating capacity additions and retirements from oil, gas, and coal power plants.

In the US Energy Information Administration (EIA) Annual Energy Outlook's Reference case, natural gas-fired combined-cycle generation capacity will continue to be added steadily through 2050. Significant retirements of electric generation capacity, mostly from coal, occur by 2025, while approximately 117 GW of new wind and solar capacity additions could occur between 2020 and 2023 [52]. This means that without investing in heat rate improvement technologies by 2025 to increase their efficiency, coal-fired generation systems must retire to comply with the affordable clean energy (ACE) rule or become stranded assets. The AEO2020 Reference case also shows that the low cost of natural gas prices significantly contributes to the retirements of coal-fired and nuclear power plants by 2025. Diverse policy efforts, notably increasing state RPS targets, net metering policies, and declining capital cost profile of solar, are expected to incentivize and accelerated its growth through 2050 by making the investment case for widespread solar energy deployment attractive to investors, particularly when utility-scale and small-scale applications are considered. **Table 1** summarizes the seven main drivers of stranding and the different aspects of stranded resources and assets.

According to the Potsdam Climate Institute, to meet the Paris Agreement of a temperature target below 2°C of global warming with the aim to limit it to 1.5°C, the global carbon budget of the total volume of CO_2 emissions permitted by 2050 is 886 $GtCO_2$ [74]. However, more than a third of this carbon budget has already been used up from burning fossil fuels, leaving a budget of around 565 $GtCO_2$. In the case of a 1.5°C temperature limit or even lower, this budget would be drastically contracted. It is for this reason that national, state and local governments must prioritize low-carbon transformations; for instance, (i) ramping up renewable energy over the next two decades, (2) switching from oil to less carbon-intensive gas [5, 15, 25, 42, 48, 55, 75–77], and (3) keeping large global deposits of

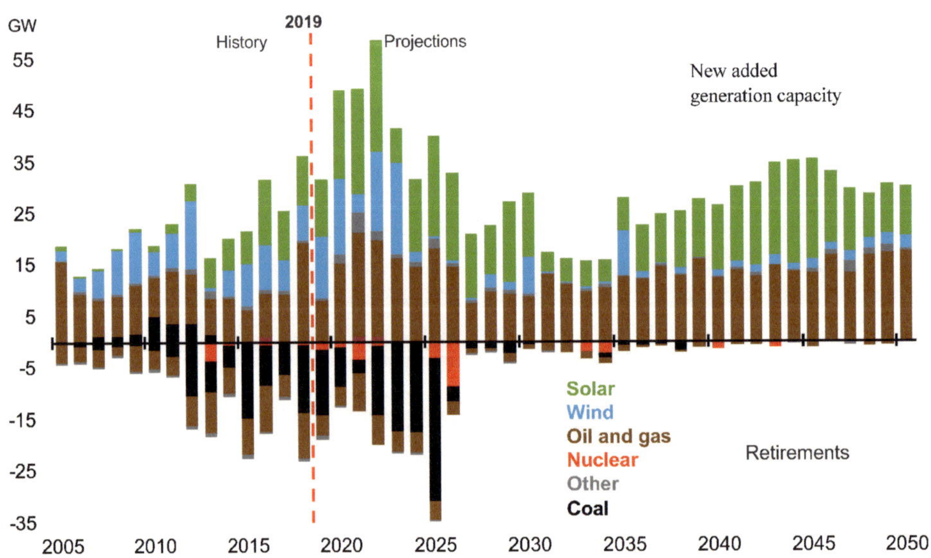

Figure 2.
Annual oil, gas, and coal electricity generating capacity additions and retirements. Source: EIA (2020).

Type of stranding	Nature of asset	Cause of stranded asset	Stranded resource	Liability	References
Economic	Viable projects receive investment (e.g., growing biofuels in deforested lands)[a]	Increased market competition affects investment in the asset (e.g., falling oil prices leads to cuts in oil exploration investments)	When it becomes uneconomical to extract/convert the resource due to low demand	Premature stranding costs (e.g., decommissioning and phase-outs costs)	[10, 53–56]
Technological	New technological breakthroughs (e.g., hydraulic fracturing, CCUS, and solar geoengineering like injecting sulfate aerosols into the stratosphere)[b]	New technologies and disruptive innovations render old technologies obsolete	Slow technological learning to access the resource (e.g., deep-sea exploitation and exploration)	Liability when technology becomes obsolete or dangerous	[57–61]
Political	The political climate is conducive for resource exploitation	Geopolitical changes like sanctions may affect assets (e.g., The Trump administration sanctions against Huawei affected Chinese oil/gas contracts)	Political strife or civil war inhibits resource exploitation	Liabilities levied against governments or organizations for (short-term) policies (e.g., aid agencies for export credits on polluting industries)	[9, 12, 62, 63]
Policy/legal	Policies and laws allow consumption, contracts, leases, and intellectual property rights/patents	New legal regime leads to asset retirement or phasing out (e.g., nuclear phase-out)	Policies or laws may restrict resource extraction or conversion (e.g., moratoria)	Pareto improvement; Liabilities for the premature stranding of investments due to policy changes (e.g., trade agreements)	[15, 64]
Spatial	The asset can be exploited	Resource depletion; water scarcity	The resource is remote (e.g., inaccessible gas or solar resource)	Liabilities for clean-up costs (e.g., Superfund clean-up costs for contaminated pollutants)	[65–67]
Social	Communities or consumers prevent the use of the asset (e.g., NIMBY ("not in my backyard") protests)	A community or consumer protests lead to its ban (e.g., Keystone Pipeline XL protests)	A community or consumers prevent the use of a resource (e.g., local fracking bans)	Compensation for resource damage (e.g., US Deepwater Horizon BP oil spill environmental damages, Nigeria's Niger Delta oil spills accidents)	[68–70]

Type of stranding	Nature of asset	Cause of stranded asset	Stranded resource	Liability	References
Ecological	Economic benefits are greater than the ecological impacts.	Ecological considerations (e.g., climate change) outweigh economic arguments.	Ecological effects inform non-use decisions of resource (e.g., large hydro dams)	Insurance or costs of adaptation borne by an investor Punitive damages incurred as injunctive relief	[56, 71–73]

[a]*Increased efficiency could create higher overall demand referred to as the Jevons paradox. For example, a shift to electric vehicle model may lead to the rebound effect, resulting in increased vehicle miles travelled, and overall rise in GHG emissions [54, 56].*
[b]*New technological breakthroughs like carbon capture, utilization and storage (CCUS) innovations can be used to extract CO_2 from power plant exhaust and industrial processes.*

Table 1.
The different aspects of stranded resources and assets.

Total proved coal reserves at the end of 2019			Total proved oil reserves at the end of 2019			Total proved gas reserves at the end of 2019		
Country	**Reserves (million tons)**	**% World**	**Country**	**Reserves (billion barrels)**	**% World**	**Country**	**Reserves (trillion cubic meters)**	**% World**
The United States	249,537	23.3%	Canada	170.8	9.8%	The Russian Federation	38.0	19.1%
The Russian Federation	162,166	15.2%	Venezuela	303.8	17.5%	Iran	32.0	16.1%
Australia	149,079	13.9%	Kazakhstan	30.0	1.7%	Qatar	24.7	12.4%
China	141,595	13.2%	The Russian Federation	107.2	6.2%	Turkmenistan	19.5	9.8%
India	105,931	9.9%	Iran	155.6	9.0%	The United States	12.9	6.5%
Indonesia	39,891	3.7%	Iraq	145.0	8.4%	China	8.4	4.2%
Germany	35,900	3.4%	Kuwait	101.5	5.8%	Venezuela	6.3	3.2%
Ukraine	34,375	3.2%	Saudi Arabia	297.7	17.1%	Saudi Arabia	6.0	3.0%
Poland	26,932	2.5%	The United Arab Emirates	97.8	5.6%	The United Arab Emirates	5.9	3.0%
Kazakhstan	25,605	2.4%	The United States	68.9	4.0%	Nigeria	5.4	2.7%
Turkey	11,525	1.1%	Libya	48.4	2.8%	Algeria	4.3	2.2%
South Africa	9,893	0.9%	Nigeria	37.0	2.1%	Iraq	3.5	1.8%
		92.8%			90.1%			84.0%

Source: BP Statistical Review of World Energy 2020 [78].
Notes: The total world proved coal, oil, and gas reserves at the end of 2019 were 1,069,636 million tons, 1735.9 billion barrels, and 198.8 trillion cubic meters, respectively. The total proved coal reserves include both anthracite and bituminous reserves and sub-bituminous and lignite reserves.

Table 2.
Global reserves of coal, oil, and gas.

coal, oil, and gas reserves "in the ground" (**Table 2**) [11, 13, 79]. This call has led to the "keep fossil fuels in the ground" initiative, "fossil fuel divestment" campaign, and "unburnable carbon" resistance movement, as a way to compel companies which are active in hydrocarbons or with high coal, oil, and gas reserves in their portfolios to reinvest elsewhere [17, 63, 79–84].

Table 2 shows the top 12 countries for each of the three fossil fuels. These coal, oil, and gas reserves represent 92.8%, 90.1%, and 84%, respectively, of the total global, proved reserves at the end of 2019 [78]. McGlade and Ekins [11] have computed a breakdown of the socially optimal distribution of stranded carbon assets that must be kept in the ground to meet the Paris Agreement temperature targets. They find that to have "a better-than-even chance of avoiding more than a 2°C temperature rise, the carbon budget between 2011 and 2050" must be kept at "around 870–1240 $GtCO_2$" p. 187. This translates to approximately one-third of global oil reserves, half of the global gas reserves, and over 80% of global coal reserves of unburnable fossil fuels. **Figure 3** summarizes the regional distribution of these unburnable reserves. These figures are in line with other estimates of the stranded coal, oil, and gas assets by other experts and organizations, that must be kept in the ground, to meet the 2°C Paris commitments [5, 10, 74, 85, 86]. However, while in the end, all carbon must be phased out, less-carbon intensive energy carriers like gas might continue to operate as a "bridging fuel" to the carbon-free economy, in tandem with renewable energy. When considering short- and long-term nature of technology rebound effects, path dependency in policymaking, and carbon lock-in in different markets [16, 20, 87–92], adopting adaptive strategies, incorporating technology transfer, and incentivizing international collaboration in energy research, are vital stratagems for managing the distributional impact of this energy transition process as well as upstream value chain requirements (such as future nuclear baseload supply and renewables-based hydrogen generation).

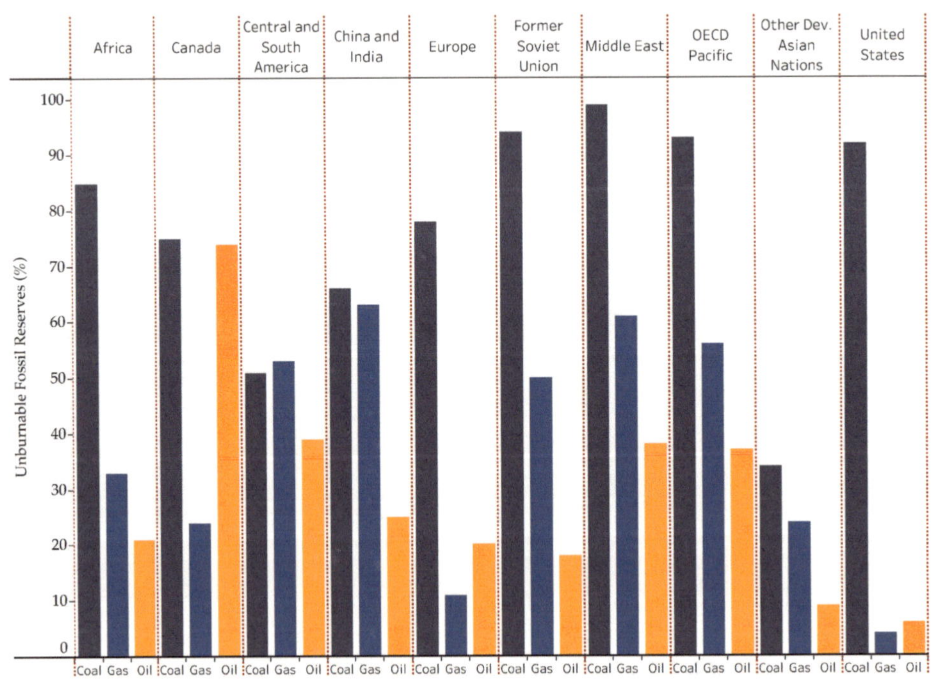

Figure 3.
Percent of regional distribution of unburnable fossil reserves before 2050 for the 2°C scenario—data from McGlade and Ekins [11], ***Table 1****.*

5. Mitigating stranded assets risks using ML and AI techniques

Returning to physical and financial carbon assets at risk of being stranded, ML and AI techniques can provide appealingly pragmatic Pareto-optimal solutions for mitigating stranding risks among different policy aspects instead of using scalarization, thereby creating a balanced transition to lower-carbon technology [93–95]. What causes assets to strand? As discussed in Section 2, multiple factors, including economic, technological like disruptive innovation, political, regulation, spatial, and societal norms, or environmental shocks, can lead to asset stranding. Stranding is not just a loss in economic value but also an irreversibility of the investments. This means that if the investments wiped out is reversible and can be adjusted for other purposes such as retooling an obsolete coal power plant to be used as a hydro generation facility; then the assets have not stranded since they can be put to different profitable use [63, 80, 86].

With respect to the unburnable carbon, stranding occurs when coal, oil, and gas companies, who have already committed heavy capital investment in related infrastructures such as exploitation, exploration, and pipelines, become hit by a sudden drop in commodity prices, leading to the stranding of their capital stocks. This could also happen when a government establishes an unanticipated Pigouvian fee, promoted by Pigou in a seminal article [96], on GHG emissions to correct for the unpriced environmental externality, either via a carbon price [97, 98] or a market-based emissions cap-and-trade mechanism [99]. This can have negative consequences for the market valuation of the upstream and downstream fossil fuel-based businesses and producers of electricity, leading to forced write-offs of their carbon assets [21] or their capital stocks getting stranded. For example, following the passage of the Powerplant and Industrial Fuel Use Act (FUA) in 1978 in response to the Arab oil embargo of 1973, a significant shortage of natural gas occurred, leading to a drop in natural gas-fired generation capacity additions. The unintended consequence of this policy-driven change in the national electricity generation mix was a shift to coal-fired generation capacity in the intervening years, leading to a rise in energy-related long-term carbon emissions.

In recent years, research shows that ML and AI are broadly powerful tools for technological progress that can be applied with a high impact in mitigating the transition to low carbon technologies, especially in tackling the problem of stranded assets in the electricity sector. Power generation and demand forecasting is one area in which ML and AI techniques can improve policy vagaries and uncertainty about future demand, thereby mitigating the risk of stranding [17]. Below are the 10 distinct opportunities for ML and AI applications in the energy sector that include:

1. Electricity scheduling and dispatch: Improving electricity scheduling and dispatch mechanisms using ML and AI tools amidst increasing variable DER generation, storage, and flexible demand.

2. Energy data analytics and informatics: Using ML supervised models, e.g., that employ regression-based techniques on cellular network data, to generate information about low-data settings and determine where electricity power lines can be placed in regions unmapped, and help improve energy access [100].

3. Energy materials research: Applying ML, AI, optimization techniques, and physics to better understand the science of energy material's crystal structure, to accelerate materials discovery for solar fuels that improves harnessing of energy from variable natural resources [101].

4. Natural gas methane detection and prevention: Employing ML and AI techniques to detect and prevent the leakage of methane from natural gas pipelines and compressor stations.

5. Nuclear fission and fusion: Application of ML and deep networks to speed up inspection of nuclear power plants and help design next-generation smart, modular nuclear reactors [102, 103].

6. Solar PV design and innovations: Using ML techniques to design controllable movable solar panels that maximize electricity production, for example, in bifacial solar modules and dual-orientation racking techniques [53, 104–107].

7. Solar PV technical and economic potential estimation: Using ML to help estimate technical and economic potential of rooftop solar PV, e.g., by optimizing Light Detection and Ranging (LIDAR)-Geographic Information Systems (GIS) imagery-rendering of size and location data for rooftop solar panels [108, 109] .

8. Wind power management and monitoring: ML-driven condition monitoring (such as dimensionality reduction algorithm like Principal Component Analysis—PCA) of wind turbine blades, including optimization of blade fault detection, power curve monitoring, and temperature monitoring [110, 111].

9. Integrated transportation planning: Using AI and ML to improve vehicle engineering, shared mobility, and shift to lower-carbon options, like rail. In the long-term, ML and AI applications can support integrated intelligent infrastructure through planning, maintenance, and operations to make transportation more efficient though the GHG reduction, provide better demand forecasts, and support smart transit policy efforts such as autonomous vehicles, alternative fuels and electrification (e.g., electric vehicles, and vehicle-to-grid algorithms), and predicting battery state and degradation rate using supervised learning techniques [112–116].

10. Urban energy planning: With ML and AI applications, available building[1] energy use data can be extrapolated to predict energy use at the city level. Furthermore, ML is uniquely capable of supporting improvements in "smart energy frameworks for smart cities" [25], including building codes, informing policymakers about utilizing urban rooftops for solar PV electricity generation [55, 108], retrofitting strategies using automated performance control [117], public-private partnerships to improve low-and moderate-income (LMI) stipulations and equitable electricity access [15, 64].

The above list is by no means exhaustive. The transformation to a low-carbon economy is occurring at an expanding rate. The technical innovations accompanying these carbon-free energy sources such as solar, wind, hydro, and geothermal energy is driving down the cost of these technologies as production increases and knowledge accumulation results from learning by doing. As a result, they are yielding substitutes for coal, oil, and progressively rendering coal, oil, and gas capital stock obsolete. It is

[1] The IPCC classifies mitigation actions in buildings into four categories: carbon efficiency (switching to low-carbon fuels or to natural refrigerants); energy efficiency (reducing energy waste through insulation, efficient appliances, better heating and ventilation, or other similar measures); system and infrastructure efficiency (e.g. passive house standards, urban planning, and district cooling and heating); and service demand reduction (behavioral and lifestyle changes).

expected that as this shift continues, new opportunities for ML and AI applications will become available, including in modeling consumer behavior and facilitating sustainable behavior change energy consumption action [3, 65, 118, 119], estimating and predicting the marginal emissions of residential energy utilization and thermal comfort in buildings in real time, on a scale of hours [57, 118], and game-theoretic modeling and design of socially beneficial energy policies like social norms, public opinions, stakeholder engagement, and education efforts [120–122]. Other breakthrough innovations might displace fossil fuels leading to stranding, and creating opportunities for ML-based electricity pricing techniques and rate design to set dynamic pricing of carbon, electricity, and consumer choice [1, 123–127], and multi-objective optimization to compute Pareto-optimal solutions for climate engineering, climate informatics, and solar geoengineering [58, 128–130]. There is a possibility that these technological innovations could create a sudden improvement in market evaluation of the renewable energy industries, while some assets of related carbon-intensive industries become stranded due to obsolescence, write-offs, or retirements.

6. Conclusion

Following the passing of the Paris Agreement on climate change, nations committed to keeping the global temperature below 2°C. Achieving this temperature target means coal, oil, and gas producers face stranding more than 80% of all these proven fossil fuel reserves and existing investments becoming stranded assets. These threats lead policy-makers and market analysts to conclude that market evaluation and capital investments of some of these carbon-intensive firms risk being stranded, unless they fundamentally change their business models per the risk of asset stranding, to cushion themselves from unanticipated economic, technological, political, regulatory, spatial, social, and environmental changes, resulting in cheap renewable substitutes for coal, oil, and gas. A pragmatic and proactive response by governments is urgently required in the form of NDCs and climate policies to guide this transition, and that puts nations on a sustained path to the 1.5 or 2°C "carbon budget." Such a process should avoid a disruptive and unorderly energy transition and macro shocks. Using ML and AI techniques to tackle the risks of stranded carbon assets and related infrastructure can enrich and inform this praxis. Stranded assets are not new in the energy sector; the physical impacts of climate change and the transition to a low-carbon economy have generally rendered redundant or obsolete electricity generation and storage assets. Low-carbon electricity systems, which come in variable and controllable forms, are essential to mitigating climate change. These systems present distinct opportunities for machine learning and artificial intelligence-powered techniques, making their applications prominent.

Sen and von Schickfus [62] calculate that €1.61 billion of security reserve or €13.38/MWh subsidy, is required to compensate coal energy assets in Germany at the risk of becoming stranded. Given the threats of sudden changes in the stringency of carbon policies and related abrupt repricing or retirement of fossil fuel assets, they also find that investors generally do care about stranded asset risk, but that they also expect to be financially compensated for stranded assets. This analysis highlights the threat of stranded asset risk in the coal industry and the need for understanding the interaction between policymaking and investors' expectations. For example, the International Renewable Energy Agency (IRENA) [131] estimates that to meet the Paris Agreement's 2°C temperature target, $1.9 trillion in electricity generation assets would be stranded after 2030. The report concludes that stranding will disproportionately affect $7 trillion in upstream energy infrastructure, of which three-quarters are in oil production. Institutional investors must tap ML and AI techniques ML to improve energy planning and system efficiency (e.g., detect and prevent the leakage of methane from natural gas pipelines, speed up inspection of nuclear power plants, and improve electricity scheduling and dispatch mechanisms). Given the vital role of the energy sector and its interrelation with the rest of the economy, using ML and AI to tackle stranded electricity assets is emerging as a cost-effective derisking strategy. Stranding and the risk of stranded carbon assets is a growing challenge requiring an interdisciplinary approach that brings together ideas from engineering, economics, and policy fields, as well as quantitative opportunities of ML, AI, optimization, and dynamical systems, to address interpretability, uncertainty quantification, and integration questions.

Author details

Joseph Nyangon[1,2]

1 Center for Energy and Environmental Policy (CEEP) at the University of Delaware, Newark, United States

2 Foundation for Renewable Energy and Environment (FREE), New York, United States

*Address all correspondence to: jnyangon@udel.edu

IntechOpen

© 2020 The Author(s). Licensee IntechOpen. Distributed under the terms of the Creative Commons Attribution - NonCommercial 4.0 License (https://creativecommons.org/licenses/by-nc/4.0/), which permits use, distribution and reproduction for non-commercial purposes, provided the original is properly cited. BY-NC

References

[1] Nyangon J, Byrne J. Diversifying electricity customer choice: REVing up the New York energy vision for polycentric innovation. In: Tsvetkov PV, editor. Energy Systems and Environment. London, UK: IntechOpen; 2018. pp. 3-23

[2] Pérez-Arriaga I, Knittel C. Utility of the Future: An MIT Energy Initiative Response to An Industry in Transition. MIT Energy Initiative: Boston, MA; 2016

[3] Taminiau J, Banks JP, Bleviss D, Byrne J. Advancing transformative sustainability: A comparative analysis of electricity service and supply innovators in the United States. 2019;**8**(4):e337

[4] IPCC. Global warming of 1.5°C, an IPCC special report on the impacts of global warming of 1.5°C above pre-industrial levels and related global greenhouse gas emission pathways, in the context of strengthening the global response to the threat of climate change, sustainable development, and efforts to eradicate poverty. In: Masson-Delmotte V, Zhai P, Pörtner HO, Roberts D, Skea J, Shukla PR, Pirani A, Moufouma-Okia W, Péan C, Pidcock R, Connors S, Matthews JBR, Chen Y, Zhou X, Gomis MI, Lonnoy E, Maycock T, Tignor M, Waterfield T, editors. Geneva, Switzerland: World Meteorological Organization; 2018

[5] Rogelj J, Schaeffer M, Meinshausen M, Knutti R, Alcamo J, Riahi K, et al. Zero emission targets as long-term global goals for climate protection. Environmental Research Letters. 2015;**10**(10):105007

[6] Riahi K, Kriegler E, Johnson N, Bertram C, den Elzen M, Eom J, et al. Locked into Copenhagen pledges—Implications of short-term emission targets for the cost and feasibility of long-term climate goals. Technological Forecasting and Social Change. 2015;**90**:8-23

[7] Verbong GPJ, Beemsterboer S, Sengers F. Smart grids or smart users? Involving users in developing a low carbon electricity economy. Energy Policy. 2013;**52**:117-125

[8] International Energy Agency. World Energy Outlook 2016. Paris, France: International Energy Agency (IEA); 2016

[9] Caldecott B, Dericks G, Tulloch DJ, Kruitwagen L, Kok I. Stranded Assets and Thermal Coal in Japan: An Analysis of Environment-Related Risk Exposure. Oxford, UK: University of Oxford's Smith School of Enterprise and the Environment; 2016

[10] Carbon Tracker Initiative. Unburnable carbon: Are the world's financial markets carrying a carbon bubble? 2011. Available from: https://carbontracker.org/reports/carbon-bubble/ [Accessed: April 21, 2020]

[11] McGlade C, Ekins P. The geographical distribution of fossil fuels unused when limiting global warming to 2°C. Nature. 2015;**517**(7533):187-190

[12] Bos K, Gupta J. Stranded assets and stranded resources: Implications for climate change mitigation and global sustainable development. Energy Research and Social Science. 2019;**56**:101215

[13] Caldecott B, Howarth N, McSharry P. Stranded Assets in Agriculture: Protecting Value from Environment-Related Risks. Oxford, England: University of Oxford's Smith School of Enterprise and the Environment; 2013

[14] Loorbach D. Transition management for sustainable development: A

prescriptive, complexity-based governance framework. Governance. 2010;**23**(1):161-183

[15] Bridge G, Bouzarovski S, Bradshaw M, Eyre N. Geographies of energy transition: Space, place and the low-carbon economy. Energy Policy. 2013;**53**:331-340

[16] Klitkou A, Bolwig S, Hansen T, Wessberg N. The role of lock-in mechanisms in transition processes: The case of energy for road transport. Environmental Innovation and Societal Transitions. 2015;**16**:22-37

[17] van der Ploeg F, Rezai A. Stranded assets in the transition to a carbon-free economy. Annual Review of Resource Economics. 2020

[18] Sovacool BK. The history and politics of energy transitions: Comparing contested views and finding common ground. In: The Political Economy of Clean Energy Transitions. GB: Oxford University Press; 2017. pp. 16-35

[19] Byrne J, Lund PD. Clean energy transition—Our urgent challenge: An editorial assay. Wiley Interdisciplinary Reviews: Energy and Environment. 2017;**6**(1)

[20] Aklin M, Urpelainen J. Political competition, path dependence, and the strategy of sustainable energy transitions. American Journal of Political Science. 2013;**57**(3):643-658

[21] Bertram C, Johnson N, Luderer G, Riahi K, Isaac M, Eom J. Carbon lock-in through capital stock inertia associated with weak near-term climate policies. Technological Forecasting and Social Change. 2015;**90**:62-72

[22] MacDonald S, Winner B, Smith L, Juillerat J, Belknap S. Bridging the rural efficiency gap: Expanding access to energy efficiency upgrades in remote and high energy cost communities. Energy Efficiency. 2019

[23] Liu Y, Yao X, Wei T. Energy efficiency gap and target setting: A study of information asymmetry between governments and industries in China. China Economic Review. 2019;**57**:101341

[24] Nyangon J. Distributed energy generation systems based on renewable energy and natural gas blending: New business models for economic incentives, electricity market design and regulatory innovation [PhD dissertation]. College of Engineering, University of Delaware; 2017

[25] Nyangon J. Smart energy frameworks for smart cities: The need for polycentrism. In: Augusto JC, editor. Handbook of Smart Cities. Switzerland: Springer; 2021. pp. 1-32

[26] Lopez-Rodriguez I, Hernandez-Tejera M, Lopez AL. Methods for the management of distributed electricity networks using software agents and market mechanisms: A survey. Electric Power Systems Research. 2016;**136**:362-369

[27] Kuiken D, Más HF. Integrating demand side management into EU electricity distribution system operation: A Dutch example. Energy Policy. 2019;**129**(June 2019):153-160

[28] Stavrakas V, Flamos A. A modular high-resolution demand-side management model to quantify benefits of demand-flexibility in the residential sector. Energy Conversion and Management. 2020;**205**:112339

[29] Shomali A, Pinkse J. The consequences of smart grids for the business model of electricity firms. Journal of Cleaner Production. 2016;**112**(Part 5):3830-3841

[30] Wang H, Wang S, Tang R. Development of grid-responsive

buildings: Opportunities, challenges, capabilities and applications of HVAC systems in non-residential buildings in providing ancillary services by fast demand responses to smart grids. Applied Energy. 2019;**250**(1):697-712

[31] Sheha M, Mohammadi K, Powell K. Solving the duck curve in a smart grid environment using a non-cooperative game theory and dynamic pricing profiles. Energy Conversion and Management. 2020;**220**:113102

[32] CAISO. What the Duck Curve Tells Us about Managing a Green Grid. California, United States: California Independent System Operator (CAISO); 2014

[33] Guidehouse Insights. Energy as a Service Overview. United States: Guidehouse Insights; 2019

[34] Stern N. The Economics of Climate Change: The Stern Review. Cambridge, UK: Cambridge University Press; 2006

[35] Zhen Z, Tian L, Ye Q. A simple estimate for the social cost of carbon. Energy Procedia. 2018;**152**:768-773

[36] Iribarren D, Martín-Gamboa M, Navas-Anguita Z, García-Gusano D, Dufour J. Influence of climate change externalities on the sustainability-oriented prioritisation of prospective energy scenarios. Energy. 2020;**196**:117179

[37] Holahan R, Kashwan P. Disentangling the rhetoric of public goods from their externalities: The case of climate engineering. Global Transitions. 2019;**1**:132-140

[38] Byrne J, Lund PD. Sustaining our common future: Transformative, timely, commons-based change is needed. Wiley Interdisciplinary Reviews: Energy and Environment. 2019;**8**(1)

[39] Tol RSJ. A social cost of carbon for (almost) every country. Energy Economics. 2019;**83**:555-566

[40] Rozenberg J, Vogt-Schilb A, Hallegatte S. Instrument choice and stranded assets in the transition to clean capital. Journal of Environmental Economics and Management. 2020;**100**:102183

[41] Byrne J, Wang Y, Lee H, Kim J. An equity- and sustainability-based policy response to global climate change. Energy Policy. 1998;**26**(4):335-343

[42] Roberts C, Geels FW, Lockwood M, Newell P, Schmitz H, Turnheim B, et al. The politics of accelerating low-carbon transitions: Towards a new research agenda. Energy Research and Social Science. 2018;**44**:304-311

[43] Rogge KS, Reichardt K. Policy mixes for sustainability transitions: An extended concept and framework for analysis. Research Policy. 2016;**45**(8):1620-1635

[44] Voyant C, Notton G, Kalogirou S, Nivet M, Paoli C, Motte F, et al. Machine learning methods for solar radiation forecasting: A review. Renewable Energy. 2017;**105**:569-582

[45] DasUK,TeyKS,SeyedmahmoudianM, Mekhilef S, Idris MYI, Van Deventer W, et al. Forecasting of photovoltaic power generation and model optimization: A review. Renewable and Sustainable Energy Reviews. 2018;**81**:912-928

[46] Wan C, Song Y, Xu Z, Yang G, Nielsen AH. Probabilistic wind power forecasting with hybrid artificial neural networks. Electric Power Components & Systems. 2016;**44**(15):1656-1668

[47] Nyangon J, Alabbas N, Agbemabiese L. Entangled systems at the energy-water-food nexus: Challenges and opportunities. In: Rao P, Patil Y, editors. Reconsidering

the Impact of Climate Change on Global Water Supply, Use, and Management United States. IGI Global; 2017. pp. 145-165

[48] Nyangon J, Byrne J, Taminiau J. Assessment of price convergence between natural gas and solar photovoltaic in the U.S. electricity market. Wiley Interdisciplinary Reviews: Energy and Environment. 2017;**6**(3):1-20

[49] Lightner EM, Widergren SE. An orderly transition to a transformed electricity system. IEEE Transactions on Smart Grid. 2010;**1**(1):3-10

[50] Bao W, Yue J, Rao Y. A deep learning framework for financial time series using stacked autoencoders and long-short term memory. PLoS One. 2017;**12**(7):1-5

[51] Gui EM, MacGill I. Typology of future clean energy communities: An exploratory structure, opportunities, and challenges. Energy Research & Social Science. 2018;**35**:94-107

[52] EIA. Annual Energy Outlook 2020 (AEO2020) Reference Case: Electricity. Washington, DC: U.S. Energy Information Administration (EIA); 2020

[53] Byrne J, Nyangon J, Hegedus S, Chajes M, Taminiau J, Ahmed N, et al. Feasibility Study of City-Scale Solar Power Plants Using Public Buildings: Case Studies of Newark and Wilmington Delaware with Early Investigations of Bifacial Solar Modules and Dual Orientation Racking as Tools for City-Scale Solar Development. Newark, DE: CEEP, University of Delaware; 2019

[54] Munyon VV, Bowen WM, Holcombe J. Vehicle fuel economy and vehicle miles traveled: An empirical investigation of Jevon's Paradox. Energy Research and Social Science. 2018;**38**:19-27

[55] Byrne J, Taminiau J. Utilizing the urban fabric as the solar power plant of the future. In: Droege P, editor. Urban Energy Transition. 2nd ed. New York City: Elsevier; 2018. pp. 31-49

[56] Sorrell S. Jevons' Paradox revisited: The evidence for backfire from improved energy efficiency. Energy Policy. 2009;**37**(4):1456-1469

[57] Ngarambe J, Yun GY, Santamouris M. The use of artificial intelligence (AI) methods in the prediction of thermal comfort in buildings: Energy implications of AI-based thermal comfort controls. Energy and Buildings. 2020;**211**:109807

[58] Tahmasebi P, Kamrava S, Bai T, Sahimi M. Machine learning in geo- and environmental sciences: From small to large scale. Advances in Water Resources. 2020;**142**:103619

[59] Rubin ES, Azevedo IML, Jaramillo P, Yeh S. A review of learning rates for electricity supply technologies. Energy Policy. 2015;**86**:198-218

[60] Nyangon J. The U.S. shale gas revolution and its implications for international energy policy. Green Monitor. 2015;**3**(1):184-190

[61] McClellan J, Keith DW, Apt J. Cost analysis of stratospheric albedo modification delivery systems. Environmental Research Letters. 2012;**7**(3):034019

[62] Sen S, von Schickfus M. Climate policy, stranded assets, and investors' expectations. Journal of Environmental Economics and Management. 2020;**100**:102277

[63] Ansari D, Holz F. Between stranded assets and green transformation: Fossil-fuel-producing developing countries towards 2055. World Development. 2020;**130**:104947

[64] Heiskanen E, Johnson M, Robinson S, Vadovics E, Saastamoinen M. Low-carbon communities as a context for individual behavioural change. Energy Policy. 2010;**38**(12):7586-7595

[65] Nyangon J, Byrne J. Spatial energy efficiency patterns in New York and implications for energy demand and the rebound effect. Energy Sources, Part B: Economics, Planning, and Policy. 2020;**15**(3):1-26

[66] Shaker RR, Altman Y, Deng C, Vaz E, Forsythe KW. Investigating urban heat island through spatial analysis of New York City streetscapes. Journal of Cleaner Production. 2019;**233**:972-992

[67] Zhou W, Wang J, Cadenasso ML. Effects of the spatial configuration of trees on urban heat mitigation: A comparative study. Remote Sensing of Environment. 2017;**195**:1-12

[68] Schallenberg-Rodriguez J. Renewable electricity support systems: Are feed-in systems taking the lead? Renewable and Sustainable Energy Reviews. 2017;**76**(September 2017):1422-1439

[69] Hultman N, Rebois D, Scholten M, Ramig C. The greenhouse impact of unconventional gas for electricity generation. Environmental Research Letters. 2011;**6**(4)

[70] Byrne J, Nyangon J, Deblauwe H, Oster C, Shin S, Xu J, et al. Measuring urban sustainability through common indicators and peer city benchmarking: Assessing sustainability assets for performance improvement and economic and environmental progress in Delaware. SSRN Online Journal. 2017. Newark, DE

[71] Lele SM, Brondízio ES, Byrne J, Mace GM, Martínez Alier J, Lupp JR. Rethinking Environmentalism: Linking Justice, Sustainability, and Diversity. Cambridge, MA: MIT Press; 2019

[72] Taminiau J, Nyangon J, Lewis AS, Byrne J. Sustainable business model innovation: Using polycentric and creative climate change governance. In: Fields Z, editor. Collective Creativity for Responsible and Sustainable Business Practice. Hershey, PA: IGI Global; 2017. pp. 140-159

[73] Kennedy CA, Stewart I, Facchini A, Cersosimo I, Mele R, Chen B, et al. Energy and material flows of megacities. PNAS. 2015;**112**(19):5985-5990

[74] Meinshausen M, Meinshausen N, Hare W, Raper SCB, Frieler K, Knutti R, et al. Greenhouse-gas emission targets for limiting global warming to 2°C. Nature. 2009;**458**(7242):1158-1162

[75] Goldthau A. Rethinking the governance of energy infrastructure: Scale, decentralization and polycentrism. Energy Research and Social Science. 2014;**1**(March 2014):134-140

[76] Byrne J, Taminiau J, Kim KN, Lee J, Seo J. Multivariate analysis of solar city economics: Impact of energy prices, policy, finance, and cost on urban photovoltaic power plant implementation. Wiley Interdisciplinary Reviews: Energy and Environment. 2017;**6**(4):1-16

[77] Verdolini E, Vona F, Popp D. Bridging the gap: Do fast reacting fossil technologies facilitate renewable energy diffusion? Energy Policy. 2018;**116**:242-256

[78] BP. The Statistical Review of World Energy 2020. 69th ed. London, UK: BP; 2020

[79] Larrea C, Murmis MR. Unburnable Carbon and Biodiversity: A Global Fund for Keeping Fossil Fuels in the Ground in Biodiversity Hotspots of Developing Countries. In: Fossil Fuel Supply and Climate Policy Conference. Quito, Ecuador; 2016

[80] Caldecott B. Introduction to special issue: Stranded assets and the

environment. Journal of Sustainable Finance & Investment. 2017;**7**(1):1-13

[81] Bullard N. Fossil Fuel Divestment: A $5 trillion Challenge. New York, NY: Bloomberg New Energy Finance (BNEF); 2014

[82] Jakob M, Hilaire J. Unburnable fossil-fuel reserves. Nature. 2015;**517**(7533):150-151

[83] Piggot G. The influence of social movements on policies that constrain fossil fuel supply. Climate Policy. 2018;**18**(7):942-954

[84] Bratman E, Brunette K, Shelly DC, Nicholson S. Justice is the goal: Divestment as climate change resistance. Journal of Environmental Studies and Sciences. 2016;**6**(4):677-690

[85] Millar RJ, Fuglestvedt JS, Friedlingstein P, Rogelj J, Grubb MJ, Matthews HD, et al. Emission budgets and pathways consistent with limiting warming to 1.5 °C. Nature Geoscience. 2017;**10**(10):741-747

[86] van der Ploeg F, Rezai A. The risk of policy tipping and stranded carbon assets. Journal of Environmental Economics and Management. 2020;**100**:102258

[87] Seebauer S. The psychology of rebound effects: Explaining energy efficiency rebound behaviours with electric vehicles and building insulation in Austria. Energy Research and Social Science. 2018;**46**:311-320

[88] Fischedick M, Byrne J, Hermwille L, Taminiau J, Luhmann H, Stelzer F, et al. Reflections on the state of climate change policy: From COP21 to cities. In: Lele S, Brondizio ES, Byrne J, Mace GM, Martinez-Alier J, editors. Rethinking Environmentalism: Linking Justice, Sustainability, and Diversity. Cambridge, MA: MIT Press; 2018. ISBN: 9780262038966

[89] Font Vivanco D, Kemp R, van der Voet E. How to deal with the rebound effect? A policy-oriented approach. Energy Policy. 2016;**94**:114-125

[90] Ivanova D, Vita G, Wood R, Lausselet C, Dumitru A, Krause K, et al. Carbon mitigation in domains of high consumer lock-in. Global Environmental Change. 2018;**52**:117-130

[91] Wei T, Zhou J, Zhang H. Rebound effect of energy intensity reduction on energy consumption. Resources, Conservation and Recycling. 2019;**144**:233-239

[92] Fan R, Luo M, Zhang P. A study on evolution of energy intensity in China with heterogeneity and rebound effect. Energy. 2016;**99**:159-169

[93] Choi D, Ahn Y, Choi DG. Multi-criteria decision analysis of electricity sector transition policy in Korea. Energy Strategy Reviews. 2020;**29**:100485

[94] Antunes CH, Henriques CO. Multi-objective optimization and multi-criteria analysis models and methods for problems in the energy sector. In: Greco S, Ehrgott M, Figueira JR, editors. Multiple Criteria Decision Analysis: State of the Art Surveys. New York, NY: Springer New York; 2016. pp. 1067-1165

[95] Wang N, Heijnen PW, Imhof PJ. A multi-actor perspective on multi-objective regional energy system planning. Energy Policy. 2020;**143**:111578

[96] Pigou AC. A special levy to discharge war debt. The Economic Journal. 1918;**28**(110):135-156

[97] Creti A, Jouvet P, Mignon V. Carbon price drivers: Phase I versus Phase II equilibrium? Energy Economics. 2012;**34**(1):327-334

[98] Chesney M, Gheyssens J, Pana AC, Taschini L. Environmental Finance

and Investments. 2nd ed. Switzerland: Springer; 2016

[99] Coase RH. The problem of social cost. In: Gopalakrishnan C, editor. Classic Papers in Natural Resource Economics. London: Palgrave Macmillan UK; 1960. pp. 87-137

[100] Gershenson D, Rohrer B, Lerner A. A new predictive model for more accurate electrical grid mapping. 2019; Available from: https://engineering.fb.com/connectivity/electrical-grid-mapping. [Accessed: May 23, 2020]

[101] Butler KT, Davies DW, Cartwright H, Isayev O, Walsh A. Machine learning for molecular and materials science. Nature. 2018;**559**(7715):547-555

[102] Chen F, Jahanshahi M. NB-CNN: Deep learning-based crack detection using convolutional neural network and naïve Bayes data fusion. IEEE Transactions on Industrial Electronics. 2018;**65**(5):4392-4400

[103] Kates-Harbeck J, Svyatkovskiy A, Tang W. Predicting disruptive instabilities in controlled fusion plasmas through deep learning. Nature. 2019;**568**(7753):526-531

[104] Tillmann P, Jäger K, Becker C. Minimising the levelised cost of electricity for bifacial solar panel arrays using Bayesian optimisation. Sustainable Energy & Fuels. 2020;**4**(1):254-264

[105] Ripalda JM, Buencuerpo J, García I. Solar cell designs by maximizing energy production based on machine learning clustering of spectral variations. Nature Communications. 2018;**9**(1):5126

[106] Bayesian optimization for maximum power point tracking in photovoltaic power plants. In: 2016 European Control Conference (ECC) Aalborg, Denmark: IEEE. 2016

[107] Kopecek R, Shoukry I, Libal J. Cost/kWh thinking and bifaciality: Two allies for low-cost PV of the future. Photovoltaics International. 2015;(4th Quarter):88-97

[108] Taminiau J, Byrne J. City-scale urban sustainability: Spatiotemporal mapping of distributed solar power for New York City. WIREs Energy and Environment. 2020:1-24. DOI: 10.1002/wene.374

[109] Walch A, Castello R, Mohajeri N, Scartezzini J. Big data mining for the estimation of hourly rooftop photovoltaic potential and its uncertainty. Applied Energy. 2020;**262**:114404

[110] Jiménez AA, García Márquez FP, Moraleda VB, Gómez Muñoz CQ. Linear and nonlinear features and machine learning for wind turbine blade ice detection and diagnosis. Renewable Energy. 2019;**132**:1034-1048

[111] Stetco A, Dinmohammadi F, Zhao X, Robu V, Flynn D, Barnes M, et al. Machine learning methods for wind turbine condition monitoring: A review. Renewable Energy. 2019;**133**:620-635

[112] Davis SJ, Lewis NS, Shaner M, Aggarwal S, Arent D, Azevedo IL, et al. Net-zero emissions energy systems. Science. 2018;**360**(6396):eaas9793

[113] Zeng W, Miwa T, Morikawa T. Application of the support vector machine and heuristic k-shortest path algorithm to determine the most eco-friendly path with a travel time constraint. Transportation Research Part D: Transport and Environment. 2017;**57**:458-473

[114] Kaack LH, Vaishnav P, Morgan MG, Azevedo IL, Rai S. Decarbonizing intraregional freight systems with a focus on modal shift. Environmental Research Letters. 2018;**13**(8)

[115] Sovacool BK, Kester J, Noel L, Zarazua de Rubens G. Are electric vehicles masculinized? Gender, identity, and environmental values in Nordic transport practices and vehicle-to-grid (V2G) preferences. Transportation Research Part D: Transport and Environment. 2019;**72**:187-202

[116] Kester J, Noel L, Lin X, Zarazua de Rubens G, Sovacool BK. The coproduction of electric mobility: Selectivity, conformity and fragmentation in the sociotechnical acceptance of vehicle-to-grid (V2G) standards. Journal of Cleaner Production. 2019;**207**:400-410

[117] Taminiau J, Sanchez CD, John B, Shin S, Xu J. Risk mitigation in energy efficiency retrofit projects using automated performance control. In: Nyangon J, Byrne J, editors. Sustainable Energy Investment: Technical, Market and Policy Innovations to Address Risk. London, UK: IntechOpen; 2020

[118] Kara EC, Roberts CM, Tabone M, Alvarez L, Callaway DS, Stewart EM. Disaggregating solar generation from feeder-level measurements. Sustainable Energy, Grids and Networks. 2018;**13**:112-121

[119] Papadopoulos S, Kontokosta CE. Grading buildings on energy performance using city benchmarking data. Applied Energy. 2019;**233-234**:244-253

[120] Safarzadeh S, Rasti-Barzoki M. A game theoretic approach for assessing residential energy-efficiency program considering rebound, consumer behavior, and government policies. Applied Energy. 2019;**233-234**:44-61

[121] Giest S, Mukherjee I. Behavioral instruments in renewable energy and the role of big data: A policy perspective. Energy Policy. 2018;**123**:360-366

[122] Zhang J, Zhang W, Song Q, Li X, Ye X, Liu Y, et al. Can energy saving policies drive firm innovation behaviors? Evidence from China. Technological Forecasting and Social Change. 2020;**154**:119953

[123] Faruqui A, Arritt K, Sergici S. The impact of AMI-enabled conservation voltage reduction on energy consumption and peak demand. The Electricity Journal. 2017;**30**(2):60-65

[124] Mowlaei ME, Saniee Abadeh M, Keshavarz H. Aspect-based sentiment analysis using adaptive aspect-based lexicons. Expert Systems with Applications. 2020;**148**:113234

[125] Faruqui A, Leyshon K. Fixed charges in electric rate design: A survey. The Electricity Journal. 2017;**30**(10):32-43

[126] Wei S, Chongchong Z, Cuiping S. Carbon pricing prediction based on wavelet transform and K-ELM optimized by bat optimization algorithm in China ETS: The case of Shanghai and Hubei carbon markets. Carbon Management. 2018;**9**(6):605-617

[127] Faruqui A, Hajos A, Hledik RM, Newell SA. Fostering economic demand response in the Midwest ISO. Energy. 2010;**35**(4):1544-1552

[128] Han J, Ahn Y, Lee I. A multi-objective optimization model for sustainable electricity generation and CO_2 mitigation (EGCM) infrastructure design considering economic profit and financial risk. Applied Energy. 2012;**95**:186-195

[129] Ma J, Cheng JCP, Jiang F, Gan VJL, Wang M, Zhai C. Real-time detection of wildfire risk caused by powerline vegetation faults using advanced machine learning techniques. Advanced Engineering Informatics. 2020;**44**:101070

[130] Jakariya M, Alam MS, Rahman MA, Ahmed S, Elahi MML, Khan AMS, et al. Assessing climate-induced agricultural vulnerable coastal communities of Bangladesh using machine learning techniques. Science of the Total Environment. 2020:140255

[131] IRENA. Perspectives for the Energy Transition: Investment Needs for a Low-Carbon Energy System. Abu Dhabi: International Renewable Energy Agency (IRENA); 2017

Chapter 3

Risk Mitigation in Energy Efficiency Retrofit Projects Using Automated Performance Control

Job Taminiau, John Byrne, Daniel Sanchez Carretero, Soojin Shin and Jing Xu

Abstract

Performance gap concerns limit investment in the building energy efficiency retrofit market. In particular, the ability of projects to deliver on promised energy savings is commonly drawn into question. Performance risk mitigation mainly occurs through energy saving performance guarantees. Contractual stipulations arrange the conditions of the guarantee, and ceteris paribus, a higher energy saving guarantee should reduce project performance risk. Therefore, methods that yield a higher energy saving guarantee could help accelerate the market. We review the ability of "smart," automated, and connected technologies to: (a) intelligently monitor and control the performance of energy-consuming devices to reduce performance variations, (b) provide additional degrees of control over the project's performance, and, by doing so, (c) motivate the energy services company (ESCO) to raise the energy saving guarantee. Our analysis finds that use of such automated performance control could significantly raise the energy saving guarantee, making projects more likely to succeed.

Keywords: derisking energy investments, energy efficiency, performance gap, energy savings guarantee, building controls, Monte Carlo analysis, monitoring and verification

1. Introduction

Energy efficiency investment is the most cost-effective pathway to reduce carbon dioxide (CO_2) emissions [1, 2]. Yet, the 2014 $5.3 billion U.S. energy efficiency market can be contrasted against an estimated $92 to $333 billion overall potential [3–5]. Nominal revenue stagnation in guaranteed savings contracts for buildings between 2011 and 2014 [4] raises concerns about a seeming incapability to successfully unlock the rest of the market.

Explicit consideration of investment barriers is required to unlock energy efficiency retrofit project investment at scale [6, 7]. In particular, a critical barrier surrounds operational performance uncertainty, widely captured in the literature under the energy saving "credibility gap" or "performance gap" [8–10]. For example, expected or experienced performance risk leads project clients to emphasize concern "about [energy service companies' (ESCO's)] guaranteed savings not being achieved, causing problems to third party financing" as a top worry [11]. In a similar vein, "uncertainty of payments based on energy savings" is seen as a key market and financial barrier according to industry experts [12].

Evaluations of performance contracting projects also find over-performance. For example, an evaluation of 8541 buildings in Greece found that realized savings exceeded expectations, on average, by 44% [13]. Similarly, review of a National Association for Energy Service Companies (NAESCO) database found that 72% of projects (517 projects) experienced greater savings than were guaranteed by the ESCO, some by as much as 50% more [14]. Analysis of a Department of Energy (DOE) Super Energy Savings Performance Contract (Super ESPC) Program found that, for the aggregate of 102 projects, the value of annual cost savings exceeded the cost savings guarantee by 19% [15].

In this regard, performance contracts raise conflicting concerns: over-performance or under-performance of the guarantee? Our research question is whether risk management under a guaranteed savings contract can be improved so as to reduce ESCO tendencies to shift project risks to other parties; and can we manage risk around the guarantee in a manner that reduces the credibility gap harbored by potential clients? Our focus is on "smart controls" as one tool to address these twin problems [16, 17].

Responsible for up to 40% of CO_2 emissions, the building sector represents an especially salient target for climate change mitigation and investment [18, 19] and there is a general consensus in the literature that building controls can improve energy saving performance profiles of energy efficiency projects [20]. Control of building operations could save up to 60% of energy consumption, with most reported savings in the 10–30% range [20, 21]. Examples of commonly operational issues that result in low performance in the building sector that "smart" controls could address include continued system operation beyond necessary hours, improper technology set-points, and inadequate economizer operations [22].

The use of automated control technology options represents an emerging paradigm for monitoring and verification of project performance that can measure and control building operations in real-time [23–25]. In general, technologies within this paradigm rely on "web-based analysis software, data acquisition hardware, and communication systems (...) to store, analyze, and display whole-building, system-level, or equipment-level energy use" and typically provide sub-hourly interval meter data with graphical and analytic capabilities and assessment [26, 27]. The use of such techniques is currently largely in the pilot stage and used primarily for program targeting and opportunity identification [24]. The available technology platforms are mostly used in commercial and industrial applications [23, 24, 28–30] but "cloud computing platform[s] for real-time energy performance [monitoring and verification are] applicable to any industry and energy conservation measure" or ECM [31]. Automated building control techniques can therefore yield actionable value by monitoring and correcting, in real-time, the energy performance profile of the building [26, 27].

We review the relationship between, on the one hand, the energy efficiency sector's currently dominant risk mitigation method in the form of energy savings guarantees and, on the other hand, smart, automated building controls. In particular, we evaluate the ability of these controls to improve the monetary value of the energy savings guarantee by reducing the project's risk profile, making energy efficiency retrofit projects more likely to succeed. To that end, Section 2.0 first covers common risk conditions associated with energy efficiency retrofit projects in the built environment. Next, Section 3.0 discusses the dynamics associated with the energy savings guarantee setting process and introduces the modeling approach used to test the interaction effects between energy saving guarantees and smart, automated building controls. Section 4.0 provides the results of the analysis.

We find that building controls represent a credible mechanism for performance risk mitigation that could motivate a significant increase in the energy savings guarantee. Section 5.0 concludes the chapter.

2. Conventional energy performance contracting risk mitigation

Inherent risks accompany energy performance contracting and clear risk allocation is critical to avoid dispute or litigation [11, 32–34]. As a prominent risk mitigation option, contractual agreements are used and, principally, performance contracts between ESCO and client can be formulated as either so-called shared savings contracts or guaranteed savings contracts.[1] Shared savings contracts allow the ESCO to take a share of the savings above a target level and, in this model, ESCOs typically provide project financing [11]. Under the guaranteed savings model, the ESCO guarantees a level of performance sufficient to pay back installation and financing costs if proposed energy conservation measures (ECMs) are implemented and monitored and verified according to protocol guidelines. When actual savings fall short of the guarantee, the ESCO compensates the shortfall to the client or otherwise makes the client whole. The ESCO does not benefit from performance levels that are above the guarantee. The ESCO market now mostly uses the guaranteed savings model [14, 35].

Under the guaranteed energy savings model, project clients are typically responsible for obtaining financing either from internal funds or from external third-party investors (e.g. a bank or financial institution) [11, 36, 37]. A range of factors can cause energy savings uncertainty, including monitoring and verification risk, financing risk, and technology risk (**Table 1**). These factors complicate risk assessment, limiting the usefulness of conventional risk screening tools such as simple payback [38].

Table 1 shows also that energy savings guarantee contract design options typically include specific stipulations for these risk categories. The actual value of the guaranteed energy savings stands out as a key contributor to the project's overall viability that is easily communicated to project client and potential third-party investor. In this, ESCOs face a double-edged incentive. On the one hand, the ESCO can be inclined to set the guarantee below the expected savings of the project to lower downside risk (e.g. lower risk of dispute with the client, lower risk of needing to deliver shortfall compensation, etc.). On the other hand, a higher guarantee

Category	Manifestation	Causes	EPC contract design
Financial	Payment default	Insufficient savings	Guaranteed savings
Technology	Equipment fault	Poor maintenance	Diagnostics
Operational	Unexpected use	Baseline changes	IPMVP[a]
Monitoring and verification	Modeling errors	Incorrect assumptions	IPMVP[a]
Economic	Fuel cost increases	Price volatility	Price escalator

For further discussion on the topic, see Refs. [11, 32, 39, 40].
[a]*International Performance Measurement and Verification Protocol.*

Table 1.
Examples of relevant risks.

[1] Other contractual agreement forms, such as first out or chauffage, are also available but are not evaluated here.

makes the ESCO more competitive and more likely to win the project bidding process. No 'rule of thumb' can be clearly identified for setting the guaranteed savings value [41] although some argue that ESCOs typically set the savings value of the guarantee *below* predicted performance using ESCO-specific risk tolerances on individual energy conservation measures [42–44].

At around 15 terawatt-hours in 2012 electricity savings, the public/institutional sector dominates the 34 terawatt-hour U.S. energy efficiency market [4, 45]. This market segment is especially relevant as clients in this sector often finance up to 100% of project costs [35]. The guaranteed savings model is compelling for public/institutional property owners as they operate in a capital deficient, maintenance-deferred environment [46]. The guarantee, supported by a creditworthy ESCO, represents a financial commitment that addresses downside risk, making it easier for these property owners to attract the capital needed for the project. However, energy savings guarantees come at a cost:

- The ESCO embeds their profit directives into the value of the guarantee.
- Contractors will not typically assume risks that cannot be managed in a direct fashion. Technologies that have a proven track record of managing additional risks, therefore, could convince the ESCO to embed the dimension into the guarantee.
- Any remaining unbounded risks (any risks not captured in the guarantee) is transferred to the other parties (i.e. the client and any potential investors).
- Savings guarantees are commonly set (well) below the achievable savings in order to build-in risk protection for the ESCO.

A benchmarking database of about 6100 projects maintained by Lawrence Berkeley National Laboratory (LBNL) in partnership with NAESCO appears to

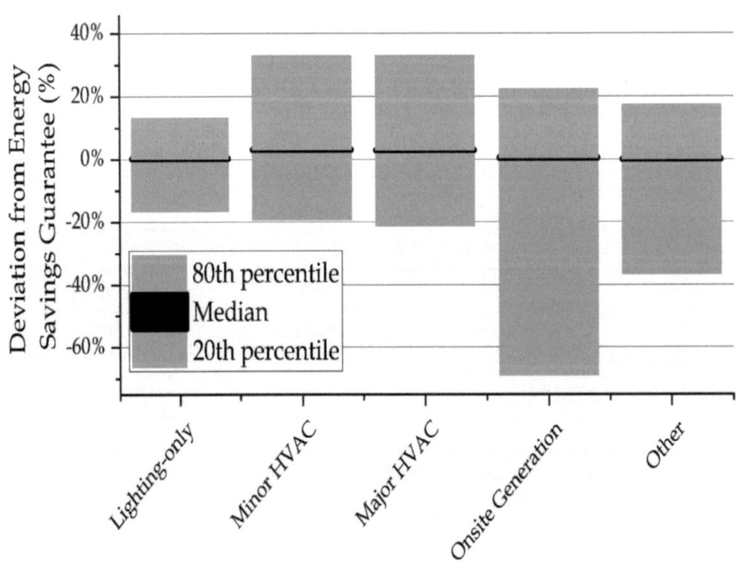

Figure 1.
Realized energy savings against guarantee at public properties (1990–2017). Note: Presented here is the 20th percentile (lower end), median (black horizontal bar), and the 80th percentile (high end) (n = 1652). Source: https://eprojectbuilder.lbl.gov/home/#/benchmark.

underscore the importance of getting the guarantee right: realized savings often deviate from the guarantee in both ways, sometimes significantly so (see **Figure 1**). Discrepancies between predicted and actual metered building energy use found in evaluations resulted in the introduction of the 'credibility gap' [8–10, 13, 47–50]. Risk mitigation in guaranteed savings projects, therefore, could deliver substantial energy performance improvements and attractive financial arrangements for, especially, public sector clients.

3. Strategic implementation of automated controls

A 2018 market analysis by a leading industry actor found that building control improvements are "the most popular investment for the next 12 months among U.S. organizations" as 68% of survey respondents indicated plans to invest in (additional) controls [51]. A 2014 estimate by McKinsey Company suggested the intelligent building control market could reach an annual $59 billion (in 2009 dollars) by 2019 [52]. Clients that have used such technology suites indicate a high level of satisfaction: 19 out of 21 cases evaluated in one assessment reported automated measurement, verification, and control as critical in achieving energy savings [27]. Trust-building and other benefits accrue from use of automated performance control options. Many of these benefits can be connected to the risks identified in **Table 1**.

3.1 Benefits of automated controls

Implementation of automated building controls could help prevent project under-performance. The potential for this technology option is substantial. For example, an assessment by the U.S. Energy Information Administration (EIA) documented in 2012 that over 85% of commercial buildings in the United States have inadequate control infrastructure in place ([53], as quoted by [54]). In addition, it has been broadly established that advanced control measures can improve performance and save 10–30% of energy consumption [20, 54–58]. For lighting, for instance, a combination of improved lighting devices and controls can reduce commercial lighting energy use by 81% [59]. A meta-analysis looking at the savings generated by lighting controls in commercial buildings by isolating the control function found savings ranging from 28 to 40% with combined operation of sophisticated controls achieving higher saving rates [60].

Whole-building energy management systems integrate a variety of end-uses (including services beyond energy such as security). A survey of zero net-energy buildings that use building controls found that 91% of the commercial buildings surveyed in North America relied on control systems that integrate multiple end-uses with 67% using a fully integrated controls architecture capable of controlling all end-uses centrally [61]. Many of these systems do often still rely on the occupant for some part of the successful operation of the controls: 74% of the buildings surveyed have integrated controls system sequences that are not fully responsible for driving performance, relying instead on the occupant [61].

Relative to the potential, significant under-adoption of the technology suites can be observed and this is often attributed to the high cost associated with whole-building applications [27]. The suite of technologies is typically deployed as software as a service (so-called SaaS) offerings, delivering capabilities on a subscription-type basis [27]. In other words, up-front expenditures for items such as licensing and system configuration are accompanied by recurring subscription fees which spread out the cost of the entire system over its lifetime. Nevertheless, up-front cost estimates range from $10 to $3400 per point with most in the $100 to $500 per point range [27]. In addition, the recurring costs range from $5 to $3100 per point [27]. Put together, 5-year ownership estimates ranged from $140 to $16,000 per point [27]. A point is a single datum that is trended, stored, and available for normalization and data analysis across use cases and comprehensive, whole-building systems can have thousands of points. For example, a use cases overview of a major controls company shows how a project involving three federal office buildings contained 18,000 points [62]. Therefore, at the median 5-year ownership costs found by Ref. [27] of $1800 per point, a fully integrated energy management system could cost as much as $32 million for the three federal office buildings.

Strategic design and selection of automated control technologies at the end-use level might overcome this barrier. Versions of partial integration deployment strategies can be observed in the market: the survey of zero net-energy buildings found that partial integration of end-uses occurred in 24% of the buildings while 9% had no whole-building controls architecture at all but, instead, used controls only at the end-use level [61]. At this level of operation, there is an expectation of significant cost reduction to the point where control technology cost can be brought down from an estimated $150–$300 per node to $1–$10 per node using low-cost, self-operated, and wirelessly connected end-use level devices [63]. Lower costs opens the door for automated controls to fulfill performance control functions for key ECMs.

Coordinated implementation of end-use level automation could enable projects to reap additional benefits:

- *Operational and engineering risk* reductions include time efficiency, improved accuracy, and possibilities for standardization and certification. For instance, automated performance control accelerates whole-building assessment from a typical 4 days to 1 day and reduces time needed for custom engineering calculations from 6 days to 1 day [64]. Automated analysis yields actionable data within the first project month [57]. Analysis of 537 projects further shows that industry standard predictive accuracy can be achieved with only 6 months of training data [65, 66]. When assessed as part of a portfolio of buildings, predictive accuracy improves further leading to the conclusion that these models are "compellingly accurate" [64]. Automated data analysis enabled by automated control enables attribution of consumption pattern variation, standardization and certification—a key need of the sector to develop investor-ready program design [7].

- *Monitoring and verification risk* reductions include portfolio level analysis, benchmarking, improved sampling, and fast anomaly and fault detection. Real-time and high-resolution automated control of performance makes even small or portfolios of projects capable of processing "big data". Further, scalability and precision allows larger sample sizes, retrieving feedback on the performance of diverse aspects of the retrofit project. Real-time data collection and control enables faster anomaly or fault detection and interface options such as online dashboards empower clients and ESCOs to mitigate under-performance.

- *Economic risk* reduction benefits from automated control include real-time utility tariff and energy consumption analysis to validate utility bills through, among others, (a) continuous monitoring and management of peak load consumption; (b) streamlining of utility-related processes to, for example, minimize personnel requirements; and (c) identification of metering or billing errors by automatically crosschecking consumption patterns with utility bills.

- *Financial risk* reduction is achieved through uncertainty mitigation and improved project finance-ability. Strategic use of automated control can deliver investor-ready program design and enhance the energy savings guarantee. Data generated by automated control can improve project finance-ability as it supports, among others, (a) accurate savings estimates, (b) risk management of operational and performance uncertainty, and (c) quick remediation of potential energy saving shortfalls.

3.2 Modeling the contribution of automated controls in the energy savings guarantee setting process for building retrofit projects

The level of savings that an ESCO will guarantee is principally influenced by: (a) the project's ability to confidently deliver savings and (b) the risk tolerance of the ESCO. A simplified interaction dynamic between project host and ESCO is provided in **Figure 2**. For a hypothetical project, **Figure 2** shows that the project's savings can exceed a low guarantee but will likely fall short when a (very) high guarantee is used. Under a guaranteed savings structure, savings above the guarantee are awarded to the project host while savings that fall short of the guarantee negatively impact the ESCO. This is illustrated in **Figure 2** by the green and red areas, respectively. From the perspective of the project host, savings that exceed the guarantee are welcome but, critically, these savings are not guaranteed and, therefore, are not available to underwrite the initial investment. However, savings that fail to reach the level of the guarantee prompt the project host to argue for compensation that

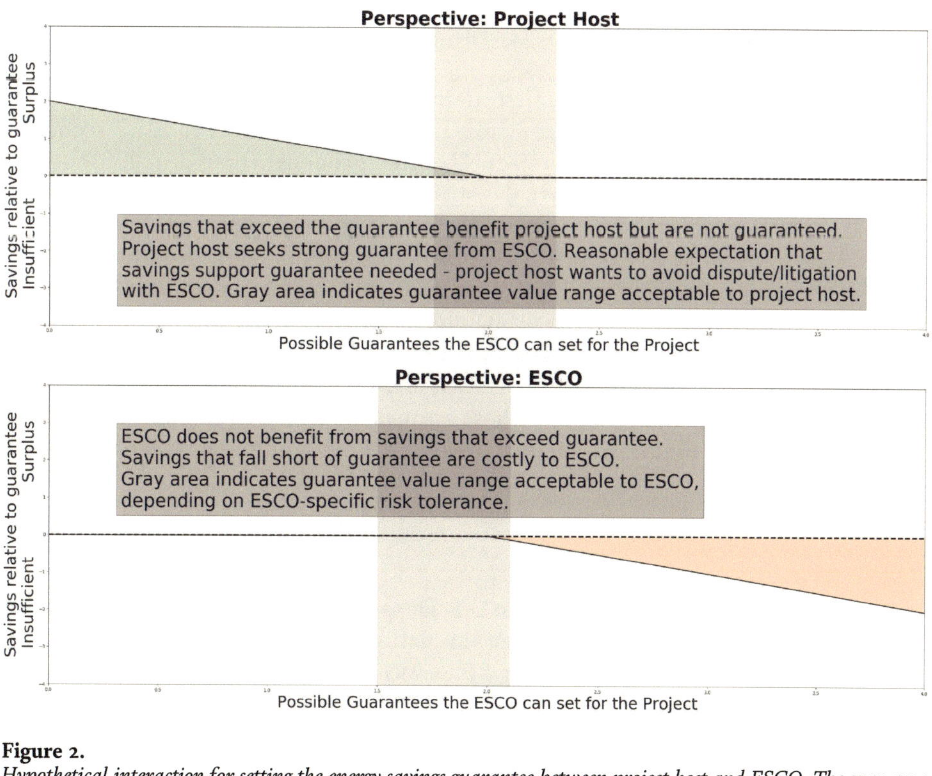

Figure 2.
Hypothetical interaction for setting the energy savings guarantee between project host and ESCO. The gray area provides an indication of agreeable guarantee levels under the modeled conditions.

could be disputed by the ESCO. This can be an arduous process. The project host, overall, is interested in a high guarantee but concerned about disputes with the ESCO [11, 12].

Surplus savings above the guarantee occur when the project over-performed relative to the guarantee. In this case, the ESCO is not at risk of claims for compensation. While this sounds appealing, it also means that the project bid by the ESCO could have been more competitive. Insufficient savings to cover the guarantee lower the overall return on the project and could even represent a net-loss for the ESCO. From the perspective of the ESCO, this should be avoided whenever possible. As illustrated in **Figure 2**, a hypothetical range of possible guarantee values that are acceptable to the ESCO can be identified. In general, the ESCO is incentivized to place the guarantee below expected savings but is hesitant to place it too low.

So far, we have established that energy efficiency savings performance can be uncertain. One way to consider this uncertainty is to reflect on energy savings performance as a stochastic distribution of possible savings. A broad distribution of energy savings i.e. a higher probability for adverse circumstances as listed in **Table 1** presents a higher risk for the ESCO that performance levels will be below the guarantee (e.g. [67]) and, ceteris paribus, is accompanied by a lower guarantee. To estimate these considerations, we follow the model proposed by Deng et al. [41]. This approach calculates, from the perspective of the ESCO, the annual and total profit for a series of possible guarantees against a Monte Carlo analysis-derived savings profile. We use the approach to calculate the guarantee level where, for a given risk tolerance, the ESCO will is unlikely to experience losses due to insufficient savings and resulting claims for compensation by the project host. In other words, consider a Monte Carlo simulation of a project's performance that results in a stochastic distribution of possible savings. In this case, a 95% risk tolerance would result in a guarantee level where 95% of all the simulated outcomes deliver savings sufficient to cover the guarantee in each year of the project lifetime. Then, the maximum guarantee within the ESCO's risk tolerance is selected as a probable guarantee for the project.

The steps of the analysis are to, first, derive possible performance profiles for pre-retrofit, post-retrofit without controls, and post-retrofit with controls scenarios for each year of the hypothetical project. This step produces three distributions of performance that approximate normal distributions. The contribution of building controls, here, is to substantially narrow the distribution of post-retrofit performance, leading to more secure savings profiles. In addition, controls improve actual building operations, leading to a savings profile with a higher overall average savings. The second step is to take the probabilistic savings profile and compare it against many possible guarantee values to identify the moments where the savings fall short. From the perspective of the ESCO, any moments where the savings exceed the guarantee are set to zero (these savings are awarded to the project host). Finally, within the stated risk tolerance of 95%, the maximum guarantee where savings are sufficient to cover the guarantee is calculated.

3.3 Software stack and data inputs

The primary software element is the U.S. Department of Energy's (DOE) Energy Plus software: a leading building energy simulation tool in the energy efficiency industry [68–70]. Advantages of Energy Plus include first-principles, text input–output work-flow that can be automated [71] and availability of benchmark building model databases (16 building types across 16 locations and three construction periods) [72, 73]. Within Energy Plus, we made use of DOEs prototypical commercial building models that describe typical building layout, geometry, energy consumption, etc. for buildings in the Delaware region constructed before 1980 [73, 74]. In particular, we model the performance of the "large office" building benchmark. The energy performance of the large office benchmark building was simulated using Energy Plus version 8.6.0. This DOE benchmark building reflects a possible building operated by the public sector, the dominant user currently of energy savings guarantee projects.

The large office benchmark building is a 46,320 square meter, 12-story office building (including basement) with total annual baseline consumption of 26,358 GJ of electricity and 7266 GJ of natural gas to fulfill its end-use functions or 725.9 MJ/m^2. Notably, over half of the buildings energy consumption serves interior lighting (9422.03 GJ or 28.1%) or interior equipment (8384 GJ or 25.2%). Heating is third most responsible for annual energy consumption (7265 GJ or 21.6%).

Possible ECMs were identified using research results from Lawrence Berkeley National Laboratory (LBNL), specifically the Commercial Building Energy Saver (CBES) project (http://cbes.lbl.gov/ and Refs. [75–77]). This ECM selection was further supported by data from the Building Component Library and several articles using a similar methodological approach [11, 36, 43, 71]. Finally, our research team had access to guaranteed energy savings agreements (GESAs) provided by ESCOs for other projects in Delaware and across the United States. Data from these GESAs was used to complete ECM profile selection by looking at buildings in those projects that share similarity with the benchmark building. Critically, based on a review of existing control literature, the selected ECMs listed in **Table 2** can be accompanied by automated controls.

Parametric evaluation of the building models was conducted using jEPlus software (version 1.7.2), an open-source parametric analysis tool specifically designed for Energy Plus [71] that provides flexible and structural analysis opportunities and smooth operations [81]. The tool has been used in similar investigations to determine sensitivity or optimize energy systems [82, 83]. This set-up enables Monte Carlo analysis for risk estimation and management of, among others, renewable energy projects, system planning, or system optimization [84, 85] and for energy efficiency projects in general and monitoring and verification efforts specifically [36, 44, 82]. Latin Hypercube Sampling (LHS) was used to run 10,000 simulations per jEPlus model run. LHS is a powerful tool that enables efficient stratification across the uncertain performance range [86]. Parametric evaluation was conducted on Amazon Web Services (AWS) architecture. The data inputs used for the

ECM & name	Energy plus parameter (Unit)	ECM costs
1. LED lighting upgrade	Lighting load (W/m^2)	\$0.63/m^2 [78][b]
2. Appliance upgrade	Plug load (W/m^2)	\$5.29/m^2 [79, 80]
3. Thermostat set-point update	Set-point in Celsius (C)	\$49.10/thermostat [80]
4. Chiller replacement	Reference COP (fraction)[a]	\$439.48/ton [80]
5. Boiler replacement	Nominal thermal efficiency (fraction)	\$34.96/MBH [80]
6. Install high-efficiency fans	Fan total efficiency (fraction)	\$0.176–\$0.390/cfm [80]
7. Water heater replacement	Heater thermal efficiency	\$20.82/gallon [80]

[a]*Coefficient of performance.*
[b]*Typical retail prices for LED packages purchased in quantities of 1000 from major commercial distributors. Using price point estimates for 2020 for cool white LED packages at 218 lumens per Watt.*

Table 2.
Selected ECMs and key parameters.

ECM	Distribution	Input	Pre-retrofit	Post-retrofit	Source
1	Normal	Value:	16.14	6.46	[82]
		σ:	0.565	0.226	
2	Normal	Value:	10.76	8.07	
		σ:	4.549	3.412	[82]
3	Triangular[a]	*Heating*			[82]
		Value 1:	21	20	
		Value 2:	15.6	14.6	
		Min/Max:	±6.52%	±6.52%	
		Cooling			
		Value 1:	24	25	
		Value 2:	26.7	27.7	
		Min/Max:	±6.52%	±6.52%	
	Normal	Value:	5.11	6.2	[82]
		σ:	0.024	0.029	
	Normal	Value:	0.76	0.95	[44]
		σ:	0.011	0.014	
	Normal	Value:	Various	0.65	σ = 5%
		σ:	0.050	0.033	
	Normal	Value:	0.8	0.95	[44]
		σ:	0.012	0.014	

[a]*Two thermostat threshold set-points for heating and two threshold set-points for cooling are included in the model.*

Table 3.
Pre- and post-retrofit performance variation inputs for the large office.

parametric evaluation are provided in **Table 3**. Controls are assumed to be able to reduce performance variation by 90%.

4. Results

The energy efficiency retrofit project of the large office benchmark building represents a total project investment of about $1.23 million. This project investment generates energy savings compared to pre-retrofit conditions. A broad range of possible energy consumption levels exists across the 10,000 simulations modeled here for both the pre-retrofit and the post-retrofit without application of performance control. In terms of energy savings, the post-retrofit scenario provides an average annual consumption level of about 24,057 GJ compared to the pre-retrofit average consumption level of 37,034 GJ an average savings of about 35% (**Figure 3**).

As provided in **Figure 3**, the savings profile is such that, under highly unfavorable circumstances, the project could have annual performance levels that are below pre-retrofit performance. In other words, in the most efficient operation of the pre-retrofit benchmark building and the most inefficient operation of the post-retrofit model, no energy savings would occur. In fact, energy consumption could be *above* the baseline in this case. The savings profile, in short, is relatively uncertain and could benefit from the inclusion of smart performance control.

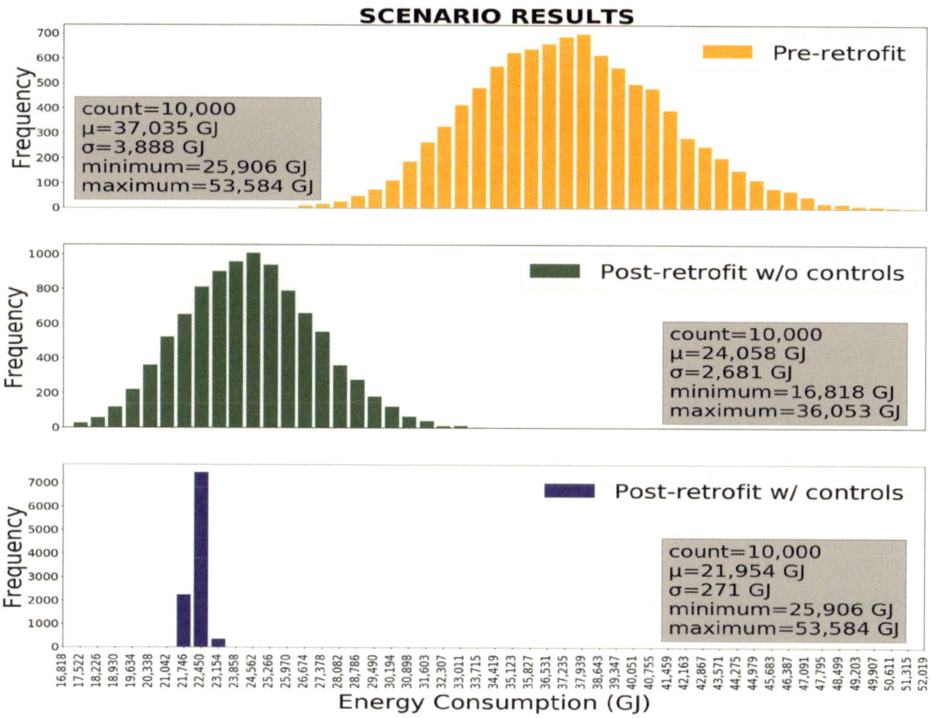

Figure 3.
Monte Carlo analysis results for energy consumption profiles in three scenarios for performance year 1 of a hypothetical project.

Figure 3 illustrates the performance profile of the post-retrofit building when performance control technologies are installed as part of the project. The controls substantially limit performance variation. In addition, the application of performance variation control also improves the overall functioning of the system, leading to higher average savings overall. For example, the average energy consumption in GJ without application of performance variation control technology is 24,058 GJ while the average energy consumption is 21,954 with application of this technology option.

The probability that a certain amount of savings in dollar terms can occur is critical in the calculus of energy savings guarantee placement. We calculate the savings distribution and probability for each year of a 20-year project using similar assumptions as Deng et al. [41] but with natural gas and electricity prices relevant for Delaware as provided by the U.S. EIA. The probability of dollar savings is influenced by the volatility and drift of natural gas prices, the contracted escalator in the price of electricity, inflation, and equipment degradation over time.

Now, we introduce the ESCO's perspective by asking the ESCO to guarantee the performance of the project. This guarantee is dependent on the energy efficiency project profile introduced above and importantly on the ESCO's tolerance for risk. We assume a highly risk averse ESCO with a risk tolerance of 95%. This means that 95% of the 10,000 simulations of the project need to exceed the guarantee in each year of the 20-year project in present value terms. The ESCO's risk for profit losses, therefore, is virtually eliminated by this risk-averse placement of the guarantee. In other words, the ESCO can reasonably expect the energy efficiency project to perform in accordance with the annual savings guarantee.

The cost savings profile of the 20-year project yields a strategic estimate of the guarantee an ESCO might be willing to provide of around $47,500. The use of

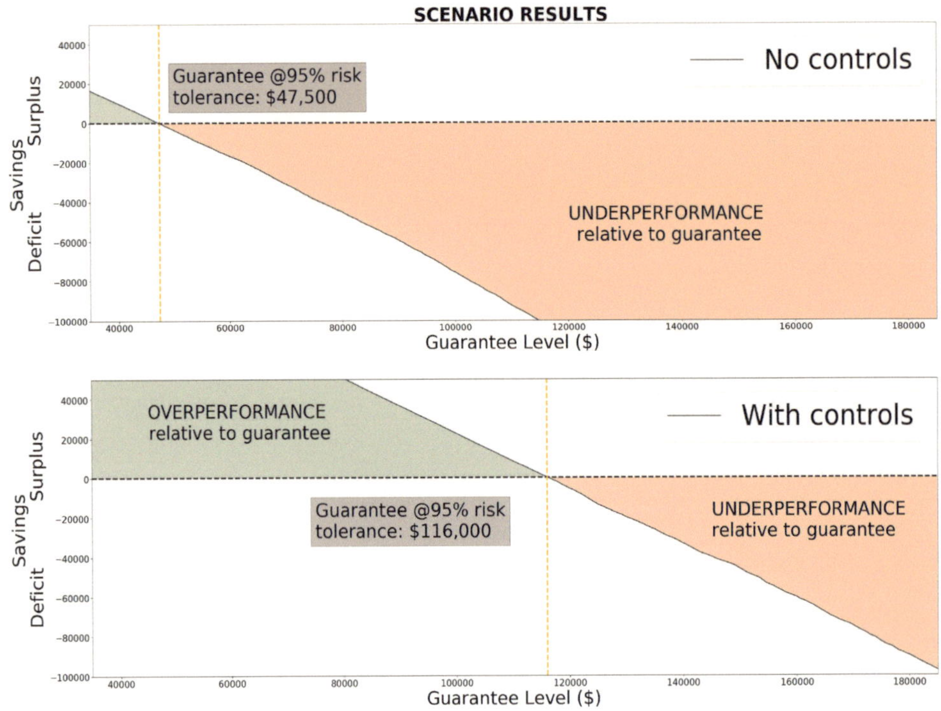

Figure 4.
Energy savings guarantee placement without and with performance control.

performance control technology is proposed as a viable risk mitigation pathway for energy efficiency performance projects. To consider the effect of using these controls, we calculate the upward movement expected in energy savings guarantee placement when the controls improve performance and reduce risk. Effectively, the scenario where controls are applied enables the ESCO to reasonably expect a higher level of performance (see **Figure 4**). The guarantee placement, at 95% risk tolerance, is increased to around $116,000.

The higher guarantee should be attractive to the project client as well as any third-party financier of the project. The ESCO, in turn, can determine the benefits of reaching this higher guarantee by weighing it against the cost of installing the additional controls. Additional research is needed to establish the precise cost profiles of the control technologies. Assuming the price points indicated by Ref [63] of $1–$10 per point are feasible for a 6000 point large office (the use case overview by a major controls company described a three-building project of large office spaces to be around 18,000 points [62]), procurement and installation of end-use and device-level control technologies could cost $6000–$60,000. Additional costs would occur from the operation of the controls. Compared to the costs associated with (a) potentially losing the project bidding process and (b) engaging in (expensive) dispute resolution regarding potential under-performance cases with the client that damages relationships and lowers future project bidding success, the installation and procurement cost of controls could be a small additional price to pay. In addition, the extra performance yields savings that can underwrite the investment, potentially leading to lower costs of financing and other costs associated with project risk.

Overall, the control function modeled here:

1. Improves overall savings by achieving a lower average consumption;

2. Lowers performance risk by achieving a significantly lower standard deviation of the savings distribution; and

3. Doubles the savings guarantee the ESCO can reasonably be expected to provide.

5. Conclusions

The conceptual and modeling approach introduced in this chapter targets performance uncertainty a dimension commonly neglected in energy savings calculations [69] despite its potential usefulness in the investment decision-making process [38, 87]. The stochastic profile of energy efficiency projects is illustrated both with and without the use of performance variation control technologies in an attempt to quantify the contribution of such advanced, real-time, high-accuracy control technology. In a sense, the use of this technology is expected to enable a more deterministic accounting of project performance through real-time and high-quality measurement that limits the stochastic range of performance. The conceptual and modeling approach benefits from automatic and interval performance measurement of a variety of devices and equipment (either at the device-level, sub-meter level, or whole-building level) which provide previously unavailable insights into the overall project [27].

The approach devised and tested in this chapter could help accelerate ongoing efforts to improve investor confidence and strengthen the energy efficiency market. For example, ongoing efforts to enhance investor confidence include the Investor Confidence Project from U.S.-based Environmental Defense Fund (EDF) or the European Energy Efficiency Financial Institutions (EEFIG) plan to compile an open source database for energy efficiency finance performance. In particular, the use of advanced, real-time, high-accuracy control technology could have consequences for the placement of the ESCO guarantee in an energy performance contract project. Raising the guarantee by reducing performance variation is one hypothesized benefit of putting controls in place. We have made an attempt at quantifying this benefit for a common building type in the United States and show that controls can potentially deliver a substantial benefit. The combined application of probabilistic performance and deterministic accounting and management transforms uncertainty into metrics legible for conventional risk management strategies such as the implementation of robust energy savings guarantees. These risk management strategies can be attractive to all involved parties, including the third-party investor.

The adoption of building and technology controls for this purpose could be encouraged by city, state, and national governments through standards and building codes, control-specific public funding thresholds and guidelines, project performance database compiling, organizing sustainable energy investment forums, actively supporting the use of energy performance contracting that includes the use of controls, or developing project development assistance centers or facilities that help guide project promoters. In addition, as is being done under the aforementioned EEFIG umbrella, underwriting toolkits can be developed to assist financial institutions in scaling up the injection of capital in the energy efficiency market [7]. Such underwriting toolkits could emphasize the critical function of controls in relation to value and risk appraisal.

Placing controls as a central component of a comprehensive and innovative approach to energy efficiency could help unlock the $1.2 trillion value in the U.S. economy that is Net Present Value (NPV) positive [88]. This value is self-financing, making it an attractive option for financial institutions if the risk profile can be clearly understood and managed. According to McKinsey researchers [88], capture of the opportunity would reduce the U.S. energy bill by 23%, positioning the U.S. well on its way to meeting previously agreed-upon climate protection targets.

Acknowledgements

This research was supported by the Delaware General Assembly.

Conflict of interest

The authors declare no conflict of interest.

Notes

Supplementary information, data, and models are available at http://freefutures.org/publications/.

Author details

Job Taminiau[1*], John Byrne[1,2], Daniel Sanchez Carretero[3], Soojin Shin[2] and Jing Xu[2]

1 Foundation for Renewable Energy and Environment (FREE), New York, NY, USA

2 University of Delaware, Newark, DE, USA

3 Center for Energy and Environmental Policy (CEEP), University of Delaware, Newark, DE, USA

*Address all correspondence to: jt@freefutures.org

IntechOpen

© 2020 The Author(s). Licensee IntechOpen. Distributed under the terms of the Creative Commons Attribution - NonCommercial 4.0 License (https://creativecommons.org/licenses/by-nc/4.0/), which permits use, distribution and reproduction for non-commercial purposes, provided the original is properly cited.

References

[1] Hoffman IM, Goldman CA, Rybka G, Leventis G, Schwartz L, Sanstad AH, et al. Estimating the cost of saving electricity through U.S. utility customer-funded energy efficiency programs. In: Energy Policy. Vol. 140. 2017. pp. 1-12. DOI: 10.1016/j.enpol.2016.12.044

[2] Molina M. The best value for America's energy dollar: A national review of the cost of utility energy efficiency programs. American Council for an Energy Efficient Economy (ACEEE). 2014. Report available at https://aceee.org/research-report/u1402

[3] Stuart E, Larsen PH, Goldman CA, Gilligan D. A method to estimate the size and remaining market potential of the U.S. ESCO (energy service company) industry. Energy. 2014;**77**: 362-371. DOI: 10.1016/j.energy.2014.09.003

[4] Stuart E, Carvallo JP, Larsen PH, Goldman CA, Gilligan D. Understanding recent market trends of the US ESCO industry. Energy Efficiency. 2018;**11**(6):1303-1324. DOI: 10.1007/s12053-018-9633-9

[5] Larsen PH, Carvallo JP, Goldman CA, Murphy S, Stuart E. Updated Estimates of the Remaining Market Potential of the U.S. ESCO Industry. Lawrence Berkeley National Laboratory (LBNL); 2017. DOI: 10.2172/1393619. Retrieved from: https://www.osti.gov/servlets/purl/1393619

[6] Parker M, Guthrie P. Crossing the energy efficiency chasm: An assessment of the barriers to institutional investment at scale, a UK perspective. Journal of Sustainable Finance Investment. 2016;**6**(1):15-37. DOI: 10.1080/20430795.2016.1159650

[7] Energy Efficiency Financial Institutions Group. Energy Efficiency—The First Fuel for the EU Economy: How to Drive New Finance for Energy Efficiency Investments. United Nations Environment Programme Finance Initiative. 2015. Retrieved from: http://www.unepfi.org/fileadmin/documents/EnergyEfficiency-Buildings_Industry_SMEs.pdf

[8] Menezes AC, Cripss A, Bouchlaghem D, Buswell R. Predicted vs. actual energy performance of non-domestic buildings: Using post-occupancy evaluation data to reduce the performance gap. Applied Energy. 2012; **97**:355-364. DOI: 10.1016/j.apenergy.2011.11.075

[9] Cali D, Osterhage T, Streblow R, Müller D. Energy performance gap in refurbished german dwellings: Lessons learned from a field test. Energy and Buildings. 2016;**127**:1146-1158. DOI: 10.1016/j.enbuild.2016.05.020

[10] Imam S, Coley DA, Walker I. The building performance gap: Are modellers literate? Building Services Engineering Research and Technology. 2017;**38**(3):351-375. DOI: 10.1177/0143624416684641

[11] Lee P, Lam PTI, Lee WL. Risks in energy performance contracting (EPC) projects. In: Energy and Buildings. Vol. 92. 2015. pp. 116-127. DOI: 10.1016/j.enbuild.2015.01.054

[12] Ghosh S, Young-Corbett D, Bhattacharjee S. Barriers to the use of ESPC in the private building sector: Perception of the A/E/C/ commune. In: Paper Presented at the 47th ASC Annual International Conference, Omaha, NE, USA. 2011. Available at: http://ascpro0.ascweb.org/archives/cd/2011/paper/CPGT299002011.pdf

[13] Balaras CA, Dascalaki, EG, Droutsa KG, Kontoyiannidis S. Empirical assessment of calculated and actual

heating energy use in hellenic residential buildings. Applied Energy. 2016;**164**:115-132. DOI:10.1016/j.apenergy.2015.11.027

[14] Hopper N, Goldmand C, McWilliams J, Birr D, Stoughton McMordie K. Public and Institutional Markets for ESCO Services: Comparing Programs, Practices, and Performance. Berkeley, CA, USA: Lawrence Berkeley National Laboratory; 2005. Available at: https://emp.lbl.gov/publications/public-and-institutional-markets-esco

[15] Shonder JA, Hughes PJ. Evaluation of the Super ESPC Program: Level 2 Recalculated Cost Savings. Oak Ridge, TN, USA: Oak Ridge National Laboratory; 2007. ORNL/TM-2007/065

[16] Byrne J, Taminiau J, Saraswat M, Mishra A. Investor risk mitigation in energy efficiency retrofit projects using automated monitoring and verification techniques. In: Technical Report prepared for the Delaware General Assembly. Newark, DE: CEEP, University of Delaware; 2018

[17] Byrne J, Taminiau J, Carretero D, Shin S, Xu J. Risk mitigation in energy efficiency retrofit projects using automated monitoring and verification techniques. In: Technical Report prepared for the Delaware General Assembly. Newark, DE: CEEP, University of Delaware; 2019

[18] Ürge-Vorsatz D, Danny Harvey LD, Mirasgedis S, Levine MD. Mitigating (CO_2) emissions from energy use in the world's buildings. Building Research Information. 2007;**35**(4):379-398. DOI: 10.1080/09613210701325883

[19] Mundaca L, Ürge-Vorsatz D. Demand-side approaches for limiting global warming to 1.5 C. Energy Efficiency. 2019;**12**(2):343-362. DOI: 10.1007/s12053-018-9722-9

[20] Fernandez N, Katipamula S, Wang W, Xie Y, Zhao M. Energy savings potential from improved building controls for the US commercial building sector. Energy Efficiency. 2018; **11**(2):392-413

[21] Mills E. Building commissioning: A golden opportunity for reducing energy costs and greenhouse gas emissions in the United States. Energy Efficiency. 2011;**4**(2):145-173. DOI: 10.1007/s12053-011-9116-8

[22] Katipamula S. Improving Commercial Building Operations thru Building Re-tuning: Meta-Analysis. Pacific Northwest National Laboratory (PNNL); 2016. Available at: https://buildingretuning.pnnl.gov/publications.stm

[23] Franconi E, Gee M, Goldberg M, Granderson J, Guiterman T, Li M, et al. The status and promise of advanced MV: An overview of MV 2.0 methods, tools, and applications. In: Report. Lawrence Berkeley National Laboratory and Rocky Mountain Institute; 2017. Available at: https://eta.lbl.gov/sites/all/files/publications/lbnl-1007125.pdf

[24] Goldberg M, Marean M, Puckett C, Godin C, Todd W, Bodmann S, et al. The changing EMV paradigm: A review of key trends and new industry developments, and their implications on current and future EMV practices. Report. Northeast Energy Efficiency Partnerships DNV GL. 2015. Available at: https://neep.org/sites/default/files/resources/NEEP-DNV%20GL%20EMV%202.0.pdf

[25] Granderson J, Piette MA, Ghatikar G. Building energy information systems: User case studies. Energy Efficiency. 2011;**4**(1):17-30. DOI: 10.1007/s12053-010-9084-4

[26] Granderson J, Touzani S, Custodio C, Sohn MD, Jump D, Fernandes S. Accuracy of automated measurement and verification (MV) techniques for energy savings in commercial buildings. Applied Energy. 2016;**173**:296-308

[27] Granderson J, Lin G. Building energy information systems: Synthesis of costs, savings, and best-practice uses. Energy Efficiency. 2016;**9**(6): 1369-1384. DOI: 10.1007/s12053-016-9428-9

[28] Crowe E, Kramer H, Effinger J. Inventory of industrial EMIS for MV applications. Northwest Energy Efficiency Alliance. 2014. Retrieved from: neea.org/resources/inventory-of-industrial-management-information-systems-emis-for-m-v-applications

[29] Erickson K. Summary of commercial whole building performance programs: Continuous energy improvement and energy management and information systems. Consortium for Energy Efficiency. 2012. Retrieved from: https://library.cee1.org/system/files/library/9235/CEE_CommBldg_WBCEIEMISProgSummPublicVersion_1May2012_0.pdf

[30] Aman S, Simmhan Y, Prasanna VK. Energy management systems: State of the art and emerging trends. IEEE Communications Magazine. 2013;**51**(1): 114-119. DOI: 10.1109/MCOM.2013.6400447

[31] Ke M, Yeh C, Su C. Cloud computing platform for real-time measurement and verification of energy performance. Applied Energy. 2017;**188**: 497-507. DOI: 10.1016/j.apenergy.2016.12.034

[32] Mills E, Kromer S, Weiss G, Mathew PA. From volatility to value: Analysing and managing financial and performance risk in energy savings projects. Energy Policy. 2006;**34**(2): 188-199. DOI: 10.1016/j.enpol.2004.08.042

[33] Jinrong H, Enyi Z. Engineering risk management planning in energy performance contracting in China. Systems Engineering Procedia. 2011;**1**: 195-205. DOI: 10.1016/j.sepro.2011.08.032

[34] Bertoldi P, Boza-Kiss B. Analysis of barriers and drivers for the development of the ESCO markets in Europe. Energy Policy. 2017;**107**: 345-355. DOI: 10.1016/j.enpol.2017.04.023

[35] Stuart E, Larsen PJ, Carvallo JP, Goldman CA, Gilligan D. U.S. Energy Service Company (ESCO) Industry: Recent Market Trends. Berkeley, CA, USA: Lawrence Berkeley National Laboratory and National Association of Energy Service Companies; 2016. Available at: https://www.naesco.org/data/industryreports/esco_recent_market_trends_30sep2016.pdf

[36] Fennell P, Ruyssevelt P, Smith AZP. Financial viability of school retrofit projects for clients and ESCOs. Building Research Information. 2016;**44**(8): 889-906. DOI: 10.1080/09613218.2015.1082779

[37] Li Y, Qiu Y, Wang YD. Explaining the contract terms of energy performance contracting in China: The importance of effective financing. Energy Economics. 2014;**45**:401-411. DOI: 10.1016/j.eneco.2014.08.009

[38] Jackson J. Promoting energy efficiency investments with risk management decision tools. Energy Policy. 2010;**38**(8): 3865-3873. DOI: 10.1016/j.enpol.2010.03.006

[39] Sorrell S. The economics of energy service contracts. Energy Policy. 2007; **35**(1):507-521. DOI: 10.1016/j.enpol.2005.12.009

[40] Olinga Z, Xia X, Ye X. A cost-effective approach to handle measurement and verification uncertainties of energy savings. Energy. 2017;**141**:1600-1609. DOI: 10.1016/j.energy.2017.11.103

[41] Deng Q, Jiang X, Cui Q, Zhang L. Strategic design of cost savings guarantee in energy performance

contracting under uncertainty. 2015. DOI: 10.1016/j.apenergy.2014.11.027

[42] Shonder J. Beyond guaranteed savings: Additional cost savings associated with ESPC projects. Oak Ridge National Laboratory, ORNL/TM-2013/108; 2013

[43] Fennell PJ. The impacts of project scale, scope and risk allocation on financial returns for clients and contractors in Energy Performance Contracts a stochastic modelling analysis [doctoral thesis (PhD)]. UCL (University College London); 2018

[44] Fennell PJ, Ruyssevelt PA, Smith AZ. Energy performance contracting—Is it time to check the small print? In: Paper presented at the 4th European Conference on Behaviour and Energy Efficiency, Coimbra, 8–9 September 2016. 2016 Available at: http://discovery.ucl.ac.uk/1542172/1/160601%20behave2016%20v4.pdf

[45] Carvallo JP, Larsen PH, Goldman CA. Estimating customer electricity and fuel savings from projects installed by the US ESCO industry. Energy Efficiency. 2015;**8**(6):1251-1261. DOI: 10.1007/s12053-015-9405-8

[46] Harcourt Brown Carey, Inc. Energy efficiency financing in California, needs and gaps. Harcourt Brown Carey, Inc. report presented to the California Public Utilities Commission, Energy Division. 2011. Retrieved from: http://www.harcourtbrown.com/energy-efficiency-financing-in-california-needs-and-gaps/

[47] Sunikka-Blank M, Galvin R. Introducing the prebound effect: The gap between performance and actual energy consumption. Building Research Information. 2012;**40**(3):260-273. DOI: 10.1080/09613218.2012.690952

[48] Galvin R. Making the rebound effect more useful for performance evaluation of thermal retrofits of existing homes: Defining the energy savings deficit and the energy performance gap. Energy and Buildings. 2014;**69**:515-524. DOI: 10.1016/j.enbuild.2013.11.004

[49] Karlsson F, Rohdin P, Persson M. Measured and predicted energy demand of a low energy building: Important aspects when using building energy simulation. Building Services Engineering Research and Technology. 2007;**28**(3):223-235

[50] Coleman P, Earni S, Williams CH. Could What that ESCO Sales Rep Said Really be True? Savings Realization Rates in ESPC Versus Bid-to-Spec Projects. Lawrence Berkeley National Laboratory; 2014. Available at: https://eta.lbl.gov/publications/could-what-esco-sales-rep-said-really

[51] Johnson Controls. 2018 energy efficiency indicator survey—United States. 2018. Retrieved from: https://www.johnsoncontrols.com/-/media/jci/insights/2018/buildings/files/eei-handout-102018-united-states–final.pdf?la=en&hash=1AF63E5E025A61E1443CEB9877961E642887CF3

[52] Booth A, Greene M, Tai H. U.S. smart grid value at stake: The $130 billion question. McKinsey Company. 2014. Available at: https://www.mckinsey.com/~/media/McKinsey/Industries/Electric%20Power%20and%20Natural%20Gas/Our%20Insights/US%20smart%20grid%20value%20at%20stake/US-smart-grid-value-at-stake.ashx

[53] Energy Information Administration (EIA). Commercial Buildings Energy Consumption Survey (CBECS) Public Use Microdata. U.S. Energy Information Administration (EIA). 2012. Retrieved from http://www.eia.gov/consumption/commercial/data/2012/index.cfm?view=microdata

[54] Fernandez N, Katipamula S, Wang W, Xie Y, Zhao M, Corbin C. Impacts of Commercial Building

Controls on Energy Savings and Peak Load Reduction. Oak Ridge, TN: U.S. Department of Energy. Report; 2017. Retrieved from: https://www.pnnl.gov/main/publications/external/technical_reports/PNNL-25985.pdf

[55] Fernandez N, Katipamula S, Wang W, Huang Y, Liu G. Energy savings modeling of standard commercial building retuning measures: Large office. Buildings. 2012. Retrieved from: https://buildingretuning.pnnl.gov/documents/pnnl_21569.pdf

[56] Fernandez N, Katipamula S, Wang W, Huang Y, Liu G. Energy savings modelling of re-tuning energy conservation measures in large office buildings. Journal of Building Performance Simulation. 2015;**8**(6): 391-407. DOI: 10.1080/19401493.2014.961032

[57] Fernandez N, Taasevigen D, Underhill RM. Success of commercial building retuning in federal buildings: Results and case studies. Journal of Architectural Engineering. 2017;**23**(1): C5016001. DOI: 10.1061/(ASCE)AE.1943-5568.0000216

[58] Azar E, Menassa CC. A Comprehensive Framework to Quantify Energy Savings Potential from Improved Operations of Commercial Building Stocks. 2014. DOI: 10.1016/j.cnpol.2013.12.031

[59] U.S. Department of Energy (DOE). Quadrennial Technology Review (QTR): An Assessment of Energy Technologies and Research Opportunities—Chapter 5: Increasing efficiency of building systems and technologies. Washington, DC. 2015. Retrieved from: https://www.energy.gov/sites/prod/files/2017/03/f34/qtr-2015-chapter5.pdf

[60] Williams A, Atkinson B, Garbesi K, Rubinstein F. A Meta-Analysis of Energy Savings from Lighting Controls in Commercial Buildings. Berkeley, CA. 2011. Retrieved from: http://efficiency.lbl.gov/sites/all/files/a_meta-analysis_of_energy_savings_from_lighting_controls_in_commercial_buildings_lbnl-5095e.pdf

[61] Miller A, Lyles M, Higgins C. Building controls for energy efficiency characteristics, energy impacts, and lessons from zero net energy. Buildings. 2016 Retrieved from: https://newbuildings.org/wp-content/uploads/2016/02/Module-4-Controls-in-ZNE-Buildings-Part-1.pdf

[62] Lynxspring. Use cases. 2018. Retrieved from: https://www.lynxspring.com/images/pdf/Lynxspring_Use_Cases.pdf

[63] Kuruganti T. Low-cost wireless sensors for building monitoring applications. 2014. Retrieved from: https://www.energy.gov/sites/prod/files/2014/10/f18/emt67_kuruganti_042414.pdf

[64] Granderson J, Touzani S, Fernandes S, Taylor C. Application of automated measurement and verification to utility energy efficiency program data. Energy and Buildings. 2017;**142**:191-199. DOI: 10.1016/j.enbuild.2017.02.040

[65] Granderson J, Touzani S, Custodio C, Sohn M, Fernandes S, Jump D. Assessment of Automated Measurement and Verification (MV) Methods. Lawrence Berkeley National Laboratory; 2015

[66] Granderson J, Price PN, Jump D, Addy N, Sohn MD. Automated measurement and verification: Performance of public domain whole-building electric baseline models. Applied Energy. 2015;**144**: 106-113. DOI: 10.1016/j.apenergy.2015.01.026

[67] Qian D, Guo J. Research on the Energy-Saving and Revenue Sharing Strategy of ESCOs under the Uncertainty of the Value of Energy

Performance Contracting Projects. 2014. DOI: 10.1016/j.enpol.2014.05.013

[68] Heo Y, Choudhary R, Augenbroe GA. Calibration of building energy models for retrofit analysis under uncertainty. Energy and Buildings. 2012;**47**:550-560. DOI: 10.1016/j.enbuild.2011.12.029

[69] Kim A, Anderson S, Haberl J. Current Industry Methods for Quantifying Energy Service Projects: Key Findings and Lessons Learned. 2016. DOI: 10.1061/(ASCE)AE.1943-5568.0000191

[70] Fumo N. A review on the basics of building energy estimation. Renewable and Sustainable Energy Reviews. 2014; **31**:53-60. DOI: 10.1016/j.rser.2013.11.040

[71] Hygh JS, DeCarolis JF, Hill DB, Ranji Ranjithan S. Multivariate Regression as an Energy Assessment Tool in Early Building Design. 2012. DOI: 10.1016/j.buildenv.2012.04.021

[72] Corgnati SP, Fabrizio E, Filippi M, Monetti V. Reference buildings for cost optimal analysis: Method of definition and application. Applied Energy. 2013; **102**:983-993. DOI: 10.1016/j.apenergy.2012.06.001

[73] Deru M, Griffith B, Torcellini P. Establishing Benchmarks for DOE Commercial Building RD and Program Evaluation. California: National Renewable Energy Laboratory; 2006

[74] Deru M, Field K, Studer D, Benne K, Griffith B, Torcellini P, et al. U.S. department of energy commercial reference building models of the national building stock. In: Technical report No. TP-5500-46861. Washington, DC: National Renewable Energy Laboratory; 2011. Retrieved from http://digitalscholarship.unlv.edu/renew_pubs/44/?utm_source=digitalscholarship.unlv.edu/renew_pubs/44&utm_medium=pdf&utm_campaign=pdfcoverpages

[75] Lee SH, Hong T, Piette MA, Taylor-Lange SC. Energy Retrofit Analysis Toolkits for Commercial Buildings: A Review. 2015. DOI: 10.1016/j.energy.2015.06.112

[76] Hong T, Piette MA, Chen Y, Lee SH, Taylor-Lange SC, Zhang R, et al. Commercial Building Energy Saver: An Energy Retrofit Analysis Toolkit2015. DOI: 10.1016/j.apenergy.2015.09.002

[77] Lee SH, Hong T, Piette MA, Sawaya G, Chen Y, Taylor-Lange SC. Accelerating the Energy Retrofit of Commercial Buildings Using a Database of Energy Efficiency Performance. 2015. DOI: 10.1016/j.energy.2015.07.107

[78] Pattison M, Bardsley N, Hansen M, Pattison L, Schober S, Stober K, et al. Solid-State Lighting 2017 Suggested Research Topics Supplement: Technology and Market Context. Washington, DC: U.S. Department of Energy; 2017

[79] Hong T, Chen Y, Lee SH, Piette MA. CityBES: A web-based platform to support city-scale building energy efficiency. Urban Computing. 2016. DOI: 10.1145/12345.67890

[80] Chen Y, Hong T, Piette MA. Automatic generation and simulation of urban building energy models based on city datasets for city-scale building retrofit analysis. 2017. DOI: 10.1016/j.apenergy.2017.07.128

[81] Park S, Norrefeldt V, Stratbuecker S, Grn G, Jang Y. Methodological approach for calibration of building energy performance simulation models applied to a common measurement and verification process. Bauphysik. 2013;**35**(4):235-241. DOI: 10.1002/bapi.201310070

[82] Lee P, Lam PTI, Lee WL, Chan EHW. Analysis of an air-cooled

chiller replacement project using a probabilistic approach for energy performance contracts. Applied Energy. 2016;**171**:415-428. DOI: 10.1016/j.apenergy.2016.03.035

[83] Ramos Ruiz G, Fernndez Bandera C. Analysis of uncertainty indices used for building envelope calibration. Applied Energy. 2017;**185**(Part 1):82-94. DOI: 10.1016/j.apenergy.2016.10.054

[84] Arnold UY. Economic risk analysis of decentralized renewable energy infrastructures A Monte Carlo simulation approach. Renewable Energy. 2015;**77**:227-239. DOI: 10.1016/j.renene.2014.11.059

[85] Pereira EJ d S, Pinho JT, Galhardo MAB, Macłdo WN. Methodology of risk analysis by Monte Carlo method applied to power generation with renewable energy. Renewable Energy. 2014;**69**:347-355. DOI: 10.1016/j.renene.2014.03.054

[86] Calleja Rodrguez G, Carrillo Andrs A, Domnguez Muoz F, Cejudo Lpez JM, Zhang Y. Uncertainties and sensitivity analysis in building energy simulation using macroparameters. 2013. DOI: 10.1016/j.enbuild.2013.08.009

[87] Walter T, Price PN, Sohn MD. Uncertainty estimation improves energy measurement and verification procedures. Applied Energy. 2014;**130**:230-236. DOI: 10.1016/j.apenergy.2014.05.030

[88] Granade HC, Creyts J, Derkach A, Farese P, Nyquist S, Ostrowski K. Unlocking energy efficiency in the U.S. economy. McKinsey Global Energy and Materials. 2009. Available from: https://www.mckinsey.com/~/media/mckinsey/dotcom/client_service/epng/pdfs/unlocking%20energy%20efficiency/us_energy_efficiency_exc_summary.ashx

Chapter 4

Assessing Renewable Energy Loan Guarantees in the United States

Ryan M. Yonk

Abstract

Conceived as an idea to push financing toward underdeveloped clean energy technology to improve the environment, promote economic growth, and produce a more secure energy supply, the Title XVII loan guarantee program has likely failed to meet these objectives. Instead, it has been used as a political tool, exposed taxpayers to unnecessary risk, diverted funding from alternative clean energy investments, and primarily benefitted large, politically connected corporations.

Keywords: alternative energy, incentives, loan guarantees, renewable energy finance

1. Introduction

Conceived as an idea to push financing towards underdeveloped clean energy technology to improve the environment, promote economic growth, and produce a more secure energy supply [1] the Title XVII loan guarantee program has likely failed to meet these objectives. Instead, it has been used as a political tool, exposed taxpayers to unnecessary risk, diverted funding from alternative clean energy investments, and primarily benefitted large, politically connected corporations.

The loan guarantee programs supported under Title XVII in general aim to provide financing to projects that would otherwise be unable to secure funding in the private market. When governments initiate loan guarantee programs, they generally target fledgling companies or struggling industries. In contrast, the Department of Energy program targets specific technologies irrespective of the company investing in them. The Loan Programs Office (LPO) offers loan guarantees under authority granted in Title XVII of the Energy Policy Act of 2005 and expanded in the American Recovery and Reinvestment Act of 2009. Loan guarantees are currently available only under Section 1703, which funds high-risk clean energy technology. While the LPO still oversees loan guarantees made under the Section 1705 program (of Solyndra fame), that program that expired in 2011 [2]. The latter program was more expansive and thus makes up the lion's share of the LPO's portfolio [3]. The LPO presides over a third program financing advanced vehicle technology, but that program utilizes direct loans rather than loan guarantees and will not be discussed in this testimony.

Government loan guarantee programs present a number of policy difficulties and the Department of Energy's program is no exception. I explore how the Department's loan guarantee program distorts markets, misdirects funds, and fails to promote truly innovative technology.

IntechOpen

2. Loan guarantee programs in general

Loan guarantee programs, offered both by governments and the private sector, are intended to close a fiduciary gap between burgeoning ideas and private investment. By promising to cover loan payments if a company fails, loan guarantors allow entrepreneurs easier access to private capital. Progenitors of government programs argue that private capital is too risk averse to properly finance whatever it is they seek to subsidize. Credit guarantees in private agreements are used to mitigate risks when individuals are considering investments, but the lender is unsure of the borrower's ability to repay the loan [4].

Not all cases in which "promising" technology fails to secure private financing can be considered justification for government intervention. The inability of high-risk projects to get private backing is a feature of a free market system, not a bug. The free market is generally good at making strategic, risk-conscious investments. Evidence from the Richmond Federal Reserve Bank indicates that loan guarantees indeed attract riskier investments and encourage entrepreneurs to overinvest [5]. This is a classic moral hazard problem; when the costs of risks are removed without a corroborating reduction in reward, entrepreneurs will take risks more flagrantly [6]. The burden of proof lies with those who claim that private financiers are indeed failing particular markets. Even then, as the aforementioned Richmond Federal Reserve study concluded, grants, direct loans, or other public financing options might be superior.

Some economists do argue that adverse selection among lenders, lender apprehension about particular technologies, industries, or geographical areas, or the existence of a credit crunch can all offer theoretical justification for loan guarantees. Loan guarantee programs, offered both by governments and the private sector, are intended to close a fiduciary gap between burgeoning ideas and private investment. By promising to cover loan payments if a company fails, loan guarantors allow entrepreneurs easier access to private capital. Progenitors of government programs argue that private capital is too risk averse to properly finance whatever it is they seek to subsidize. Credit guarantees in private agreements are used to mitigate risks when individuals are considering investments, but the lender is unsure of the borrower's ability to repay the loan [4].

Not all cases in which "promising" technology fails to secure private financing can be considered justification for government intervention. The inability of high-risk projects to get private backing is a feature of a free market system, not a bug. The free market is generally good at making strategic, risk-conscious investments. Evidence from the Richmond Federal Reserve Bank indicates that loan guarantees indeed attract riskier investments and encourage entrepreneurs to overinvest [5]. This is a classic moral hazard problem; when the costs of risks are removed without a corroborating reduction in reward, entrepreneurs will take risks more flagrantly [6]. The burden of proof lies with those who claim that private financiers are indeed failing particular markets. Even then, as the aforementioned Richmond Federal Reserve study concluded, grants, direct loans, or other public financing options might be superior.

Some economists do argue that adverse selection among lenders, lender apprehension about particular technologies, industries, or geographical areas, or the existence of a credit crunch can all offer theoretical justification for loan guarantees.

Still others attest that clean energy technologies ought to be subsidized by the government because they provide social benefits in excess of what can be returned to lenders, prompting private markets to underinvest. While clean energy technology does not create any positive externalities per se, it does crowd out carbon-emitting sources of energy and therefore may counteract a negative externality. Of course, there are more direct and efficient ways of targeting the carbon problem, but subsidizing clean energy is often taken as a politically viable next best alternative [4].

3. History and background

If there is one reason to be skeptical of loan guarantee programs in general, it is the paucity of conclusive academic research on their effectiveness. In my review of the academic literature it became glaringly obvious that there is still much important research to be performed on the questions of the loan guarantee program's effects, its costs and benefits, and best program design [7–13]. Data that is exact enough to make meaningful conclusions is difficult to collect. Studies are often too specific, meaning they examine one particular program and may not provide generalizable results, or too broad to have enough data to employ proper statistical analyses. This problem is further compounded by the many types of loan guarantee programs. Some provide funding for businesses to start-up, others guarantee business expansions, and in the program in question today, encourage the use of certain technologies.

As illustrative examples, here is what preliminary economics research has said about some international forays into loan guarantees. A French program targeting new firms was said to have no impact on the total number of companies, to increase their average size, and significantly increase their risk of default [14]. An investigation into a Malaysian small and medium sized enterprise program claims "there is sufficient evidence that the Scheme has failed to meet all [its] objectives" [15].

4. Policy issues in the loan guarantee program

The loan guarantee program is well-intentioned, as most policy is, but its designers failed to fully consider many unseen effects. The US Department of Energy's program has deterred investment in other areas and made it more difficult for some to receive private investments, been used as a political tool, encouraged malinvestment, and primarily benefitted established companies with plenty of preexisting access to capital for research and development.

One key insight from policy analysis is that we must measure what matters. In the case of loan guarantee programs, simply because the program expands entrepreneurs' access to credit does not make the program a success. There are other important aspects that must be considered. Government action is not justified merely because there is a market failure. Government ought to act to fix market failures only when the net gains from resolving those problems, given the possibility of government failure, are positive. As Professor and governor of the Central Bank of Ireland Patrick Honohan writes, "With many competing pressures for public funds, an economically coherent argument in favor of a subsidized credit guarantee system needs to go a lot further than the observation that such a scheme would increase availability of credit" [4].

Federal loan guarantees can only be said to serve a public benefit if they accomplish what economists call additionality, meaning the program must be offering loans to projects that would not have otherwise garnered funding in the open market. A program that extends government assistance to projects and companies that would have no trouble securing private financing accomplishes little, adds unnecessary administrative costs, and puts taxpayer money at risk.

Some exploratory research on the additionality of loan guarantee programs for energy technology from both the DOE and USDA reveals poor additionality [16]. The early evidence suggests few loans are extended that would not otherwise be attained. Given the size and robust access to financing of many companies seeking Title XVII funding, which I will discuss momentarily, poor additionality should come as no surprise.

Even if government loans managed to accomplish perfect additionality, this alone would not be sufficient justification for the continuation of a program. Many conceive of loan guarantee programs as marginally shifting the risk calculus for private investment. In other words, guarantees allow projects that would previously have been considered barely too risky to finance to get funding. Realistically, loan guarantees completely shift the entire calculation of private investors. Securing a government loan guarantee proves to be a highly political process. Private capital often follows public capital. Despite that statement's appealing tenor, this is not a positive outcome. It means only the politically connected are funded and the extent of that problem is compounded beyond the bare dollar value of the government program.

The source of problems with government support for particular energy sources is that corporations and interest groups subvert the program to serve their private interests. Funding is allocated by political processes instead of the free choice of individuals who judge it to be a worthwhile investment. The fundamental problem at the heart of the Solyndra scandal, for example, was not that the business failed after securing a loan guarantee. After all, some failure will arise out of any loan guarantee program. Rather, the evidence that emerged following that failure demonstrated that Solyndra's path to securing a government loan guarantee had been dictated by political pressure, not market viability. As documented in a chapter of *Nature Unbound*, Solyndra's application rushed through or even skipped critical oversight steps in order to reach approval before a California trip President Obama had planned. Even when failure was imminent, personnel at the Department of Energy urged even more funding to be pumped into Solyndra in an attempt to save face, despite warning from the OMB [17].

The 2015 Inspector General's report on Solyndra confirmed that "the Department missed opportunities to detect and resolve indicators that portions of the data provided by Solyndra were unreliable" and that employees "felt tremendous pressure, in general, to process loan guarantee applications [...] based on the significant interest in the program from Department leadership, the Administration, Congress, and the applicants" [18]. Solyndra shed light on this malfeasance, but political interference is a structural problem with loan guarantee programs, not merely the fault of a single public officer, agency, or administration.

One point that is too often underemphasized is that this argument against government interference applies equally to subsidizing fossil fuels. When President Carter's administration pushed for energy independence it meant government support for coal companies along with the research funding for and promotion of renewables [19, 20]. These are at least equally problematic, and considering their size, perhaps even more so.

Most Section 1705 funding has gone to large corporations who already have access to capital for investments in research, development, and deployment. Recipients of LPO guarantees include multiple Fortune 200 companies, utility companies, and multinationals. Many are wholly owned by yet larger companies [21]. The application process itself all but ensures that only large, established companies will be capable of participating in the program. Applicants can expect to pay between $150,000 and $400,000 in fees before even being considered [22].

The full ramifications of supporting mainly large corporations are rarely understood. It does not simply mean that large corporations make risky investments and leave taxpayers to pick up the tab, but the fundamental problem is that it makes it more difficult for new ideas to emerge since it further entrenches established ideas. Research on new energy technology has stalled at least in part because of government's involvement. Government support, as a previous chief marketing officer at Tesla Motors complained, may make it easier for those who receive support, but it also makes it more difficult for new ideas to gain private funding and grow [23].

Loan guarantee programs, like any subsidy, move resources towards the subsidized good. A subsidy redirects private capital towards the subsidy because it lowers the risk and changes the risk calculation investors go through. In general, the subsidized industries see growth and investment. The unsubsidized, however, see lower investment. The subsidy distorts the market signals of profit and loss to appear as if the subsidized industries provide more value than they do.

The net result of loan guarantee programs is likely a loss in meaningful innovation. This is the fundamental problem with loan guarantees. Even if the additionality was 100 percent, the program employs poor methodology to pick those to subsidize. Political power and lobbying prowess, not the collective intelligence of all individuals in the market, allocate the funding of these programs. My analysis indicates that the unseen costs are much greater than anticipated. To some extent this position rests on a counterfactual--how do you measure what did not happen? The question of what could have been, the opportunity cost of these loans, is a serious consideration even if it is a difficult empirical one.

5. Conclusions

Preliminary examinations on the Department of Energy and USDA's programs have been discouraging, though the entire literature pleads for more concerted research efforts. The political problems associated with the funding justify further skepticism towards Section 1705 and Section 1703, as do the characteristics of their recipients.

The primary take away from my analysis is that government's attempt to promote innovation has likely done exactly the opposite. In place of these programs government would do better to simply step out of the way of entrepreneurs and individuals. As the development of the technology industry demonstrates, allowing experimentation and markets to drive innovation is a promising avenue for improving the world. In contrast to policymakers propensity to want to plan for every contingency, permissionless innovation, an idea developed by the Adam Thierer, is more likely to provide the new ideas needed to solve energy and environmental issues [24]. It calls for government officials to clear a path for entrepreneurial experimentation unfettered by precautionary regulation.

A policy of permissionless innovation is more likely to find successful solutions to the pressing environmental and energy questions, such as the potential dangers from climate change and the health issues caused by pollution, than government bureaucrats choosing projects to fund based on political considerations.

Conflict of interest

The authors declare no conflict of interest.

Author details

Ryan M. Yonk
Department of Agribusiness and Applied Economics, Center for the Study of Public Choice and Political Economy, North Dakota State University, USA

*Address all correspondence to: ryan.yonk@ndsu.edu

IntechOpen

© 2020 The Author(s). Licensee IntechOpen. Distributed under the terms of the Creative Commons Attribution - NonCommercial 4.0 License (https://creativecommons.org/licenses/by-nc/4.0/), which permits use, distribution and reproduction for non-commercial purposes, provided the original is properly cited. (cc) BY-NC

References

[1] Federal Register. Loan guarantees for projects that employ innovative technologies. 10 CFR Part 609. Vol. 74, No. 232. Pg. 63544. Retrieved from: https://energy.gov/sites/prod/files/2014/03/f14/FR-1703-Dec4.pdf

[2] Brown P. Loan Guarantees for Clean Energy Technologies: Goals, Concerns, and Policy Options. Congressional Research Service; 2012. Retrieved from: https://fas.org/sgp/crs/misc/R42152.pdf

[3] Government Accountability Office. DOE LOAN PROGRAMS: Current Estimated Net Costs Include $2.2 Billion in Credit Subsidy, Plus Administrative Expenses. 2015. Retrieved from: http://www.gao.gov/assets/670/669847.pdf

[4] Honohan P. Partial credit guarantees: Principles and practice. Journal of Financial Stability. 2009;**6**(2010):1-9

[5] Li W. Government loan, guarantee, and grant programs: An evaluation. Economic Quarterly. 1998;**84**(1998):25-51

[6] Vogel RC, Adams DW. The benefits and costs of loan guarantee programs. The Financier. 1996;**4** (1 & 2):22-29. Retrieved from: https://www.microfinancegateway.org/sites/default/files/mfg-en-paper-the-benefits-and-costs-of-loan-guarantee-programs-1996.pdf

[7] Bartik T, Bingham R. Can economic development programs be evaluated? In: Upjohn Institute Staff Working Paper 29. 1995

[8] Boocock G, Shariff MNM. Measuring the effectiveness of credit guarantee schemes: Evidence from Malaysia. International Small Business Journal. 2005;**23**(4):427-452

[9] Cressy R. Funding gaps: A symposium. Economic Journal. 2002;**112**

[10] Green A. Credit guarantee schemes for small enterprises: An effective instrument to promote private sector-led growth? In: SME Technical Working Paper. Vienna: UNIDO; 2003. p. 10

[11] Levitsky J. Credit guarantee schemes for SMEs: An international review. Small Enterprise Development. 1997;**8**(2)

[12] O'Bryan III, William E. An analysis of small business loan guarantee funds [thesis]. 2010. Retrieved from: http://digitalcommons.unl.edu/cgi/viewcontent.cgi?article=1003&context=arch_crp_theses;

[13] Cowling M, Mitchell P. Is the small firms loan guarantee scheme hazardous for banks or helpful to small business? Small Business Economics. 2003;**21**:63-71

[14] Lelarge C, Sraer D, Thesmar D. Entrepreneurship and credit constraints: Evidence from a French loan guarantee program. In: Lerner J, Schoar A, editors. International Differences in Entrepreneurship. National Bureau of Economic Research (NBER); 2010. Retrieved from: http://www.nber.org/chapters/c8218.pdf

[15] Boocock G, Shariff MNM. Measuring the effectiveness of credit guarantee schemes evidence from Malaysia. International Small Business Journal. 2005. DOI: 10.1177/0266242605054054

[16] Juchau C, Solan D. Energy technology loan guarantee programs: The search for additionality in support of commercialization. In: Working Paper from WPSA. Energy Policy Institute; 2014. Retrieved from: https://wpsa.research.pdx.edu/papers/docs/WPSA%202014%20-%20Energy%20Technology%20Loan%20Guarantee%20Programs.pdf

[17] Yonk R, Simmons RT, Sims K. Nature Unbound: Bureaucracy vs. the Environment. Independent Institute; 2016

[18] U.S. Department of Energy Office of Inspector General. The Department of Energy's Loan Guarantee to Solyndra, Inc. OIG Case No. 11-0078-I. 2015. Retrieved from: https://energy.gov/sites/prod/files/2015/08/f26/11-0078-I.pdf

[19] Carter J. The President's Proposed Energy Policy. Vital Speeches of the Day, Vol. XXXXIII, No. 14, May 1, 1977; 1977. pp. 418-420. Retrieved from: http://www.pbs.org/wgbh/americanexperience/features/primary-resources/carter-energy/

[20] Carter J. National Energy Program Fact Sheet on the President's Program. 1977. Retrieved from: http://www.presidency.ucsb.edu/ws/?pid=7373

[21] de Rugy V. Assessing the Department of Energy Loan Guarantee Program. Testimony before the House Committee on Oversight and Government Reform; 2012. Retrieved from: https://www.mercatus.org/publication/assessing-department-energy-loan-guarantee-program

[22] Loans Programs Office. Title XVII Application Process. Department of Energy; 2017. Retrieved from; https://energy.gov/lpo/title-xvii-application-process

[23] Siry D. In Role as Kingmaker, the Energy Department Stifles Innovation. Wired; 2009. Retrieved from: https://www.wired.com/2009/12/doe-loans-stifle-innovation/

[24] Thierer A. Permissionless Innovation: The Continuing Case for Comprehensive Technological Freedom. Mercatus Center; 2016. Retrieved from: https://www.mercatus.org/publication/permissionless-innovation-continuing-case-comprehensive-technological-freedom

Section 3

Energy Innovation Agendas

Chapter 5

Innovative Circular Business Models: A Case from the Italian Fashion Industry

Marco Tortora and Giuseppe Tortora

Abstract

Transition to a sustainable economy signed by a circular vision and culture asks firms for huge investments to innovate their own management, strategies, business models, products, and marketing approaches. The Agenda 2030 and the 17 Sustainable Development Goals (SDG) are an important framework for businesses to change their approach and contribute positively to the global movement to fight climate change. The question is what and how micro, small, and medium enterprises (MSMES) can contribute to reduce their impacts while creating more value for them and their stakeholders. This paper aims to answer to this question presenting a case study from Italy where an artisan small firm is innovating to create more positive impacts in circular terms. The focus will be on circular economy and the firms' material and energy strategies. In doing so, the paper will try to answer the following questions: how easy is for micro and small firms to apply circular economy strategies to contribute to reduce their environmental impacts? Does their strategy coherently compose energy and material flows? The case study will refer to the fashion system in Italy.

Keywords: circular economy, renewable energy, material flows, sustainable business model, innovation

1. Introduction

In Europe, according to the EU Commission [1, 2], the textiles and clothing sector is a strategic and important manufacturing industry that employs 1.7 million people for a turnover of EUR 166 billion. The European Union (EU) is the second largest exporter in the world for textiles and fashion. Like all the other sectors in Europe, it has been affected by many structural changes deriving from a dynamic combination of external forces, especially technological (big data, virtual reality, artificial intelligence, and blockchain), economic, and political, as well as environmental (the case of the circular economy) ones.

By definition, the fashion industry—and first of all the textile and clothing industries—refers to activities covering different steps in the production and value chain: from the transformation of natural or synthetic material into yarns and fabrics to the production of products such as synthetic yarns, bed linens, industrial filter (textiles), and clothing (from finished fabric to clothes, technical and home textiles, retail, and B2B).

IntechOpen

According to the latest data [1], the European fashion industry was composed of around 180,000 firms, accounting for a 3% share of value added and a 6% share of employment in total manufacturing. The main actors of the sector are micro, small, and medium enterprises (MSMEs), a wide group of companies with less than 10 employees that covers 99% of the workforce and produce almost 60% of the value added [3].

The changes and challenges of the latest decade have driven these companies to focus on the production of higher value-added products and niche markets were competition is not on price but quality, research, and innovation. In fact, as stated by the EU Commission [1], European producers are world leaders in the markets for technical/industrial textiles and high-quality garments with a high design content. This shift from mass low-quality and value-added productions to high-quality and value-added new designed solutions requires continuous investments and efforts to maintain its global competitiveness. Political choices and solutions should support this industry to maintain and then increase its positive tension toward the future. European policies and plans should guide sectorial policies keeping the wider vision and strategy linked to the needs and problems of firms while helping these actors to create new value balancing economic, social, and ecological impacts.

From the environmental point of view, the impact of the fashion industry is huge. The apparel sector alone—that globally reached USD 1685 billion [4]—because of its (still) linear model (fragmented and relatively low-tech supply chain processes and systems), produces major environmental impacts such as the use of large quantities of water and chemicals, high emissions of greenhouse gases, and generation of waste [5].

The Ellen MacArthur Foundation (EMF) [6] has recently assessed the global impact of the textile. The global annual data registered total greenhouse gas emissions (GHG) for a value around 1200 million tons; for water usage, 93 billion cubic meters; fertilizers for cotton, 8 million tons; pesticides for cotton, 200,000 tons, and chemicals for 42 million tons, and dyestuffs for 1 million tons.

The same Ellen MacArthur Foundation [7] advocates for the "circular economy" as the sustainable solution to reduce the negative externalities in the environment. The circular economy is an economic system [7–10], in which materials continue to circulate through continuous phases of reuse and recycle to keep value within the system as much as possible. There are many benefits of a transition to the circular economy: the economic ones are accounted for around EUR 1.8 trillion by 2030 [11], while the socioeconomic benefits are expressed in terms of enhanced energy efficiency, reduced carbon emissions, and the creation of employment in the EU [12]. According to the EC [13], the benefits for a textile sector transited to a full circular economic system should bring benefits such as a reduction in resource needs while creating new jobs due to the rise or increase to relatively new phases in the production process (collecting, sorting, and recycling of clothing). According to some studies, these socioeconomic benefits yet miss a robust available evidence [14].

By reviewing the literature from academia and society, this paper aims to focus on the challenges MSMEs (especially micro and small firms) of the fashion industry (primarily clothing) have to deal with, manage, and overcome to contribute to climate change issues while increasing their value added and/or shared value with their local communities and stakeholders. The novelty of the paper resides on the presented case study, unique in Italy by now, and the fact that it clarifies the content of these challenges—or future and potential socioeconomic benefits—and points out what micro and small firms need to contribute to sustainable development.

2. Literature

At the beginning of May 2019, the European Commission has issued five main EU policy recommendations for its 2019–2024 strategic agenda in order to give future indications for the next direction of the new EU system; the emanation of the elections was held at the end of May 2019. As usual, the EU policy recommendations are based on the results obtained by the European system in the previous policy period (i.e., 2014–2019). The recommended main policy priorities are five and named as protective Europe, competitive Europe, Fair Europe, Sustainable Europe, and Influential Europe. For the aims of this chapter, the most interesting is "Sustainable Europe" that regards the way the EU supports and promotes the embracement of the circular economy paradigm through sustainable consumption practices and the minimization and the eventual reversion of the damages caused by climate change. Connected to the idea of a Sustainable Europe, there are also the Fair Europe policy for which the EU should promote social inclusion and equality of opportunity under the European Pillar of Social Rights and "Competitive Europe" that is about the support to increasing researches and investments for environmental and technological developments.

The idea of the circular economy has a strategic and necessary choice for the sustainable development of a smart and competitive Europe, which started in December 2015 when the European Commission adopted the Circular Economy Package. [15, 16] The program was ratified by the Commission in 2014 [17], and then the newly appointed Commission withdrew the program and reintroduced and accepted it after revisions in December 2015. The package dictates the transition from a linear to a circular economic model to the members of the EU. The necessity of the transition to the circular economy system is supported by the related and estimated economic data: EU savings of around €600 billion by 2030 and the creation of more than 170,000 new jobs. These and other goals have been reported in March 2019 [18–21] when the European Commission adopted a comprehensive report on the implementation of the Circular Economy Action Plan. In fact, in the report there are the main achievements reached by the Action Plan in the last years and the future challenges.

It is clear that the EU recognizes a strategic plan and central to its development a fair and inclusive economic system based on research and investments guided by the new paradigm of the circular economy at the macro level and new business models, organizational systems, and products at the micro level. Thus, in order to understand and discuss the case study presented in the following paragraph, this part focuses on official documents and academic papers that define and describe the circular economy and the way the circular economy is changing the industry of fashion and textiles.

2.1 The academic contribution to the development and definition of the circular economy

The origins of the term circular economy go back in time till the 1960s, when Boulding [22] started discussing the limits of our planet (biophysical limits) and introduced important concepts such as closed systems, the total capital stock, the reproduction of limited stocks of inputs (material, energy), and the importance of waste recycling. Later, in the 1990s, Pearce and Turner [23], applying the first and second law of thermodynamics, developed and proposed a new economic model introducing for the first time the term circular economy. Afterwards, a growing body of literature from various disciplines has emerged [24, 25].

The first is a system approach whose aim is to transform the industrial material flows investing on design and knowledge: this is the cradle-to-cradle (C2C) design theory and standard. According to Braungart et al. [26], it is fundamental to overcome the traditional linear production system based on material flows and find a different way from "traditional" sustainability or eco-efficiency concepts. The latest ones are limited since the risk is to push the system toward an extreme dematerialization at the expenses of innovation and economic growth. The C2C approach on the contrary is based on the concept of eco-effectiveness, whose main element is design. In fact, design and innovation are key to create, maintain, and increase value, quality, and productivity within the production process and economic system. Value and quality of materials depend on the quality of knowledge and information flows that spread among all the actors in the value chains [27, 28]. The strategic role of design and innovation for circular economy is not unique to the C2C approach and standard. In fact, the concept and term circular economy typify the modern idea of regenerative and responsible product design that originates from many different whole system design concepts and authors such as Walter R Sthal [29], Gunter Pauli (Blue Economy [30]), and Karl-Henrik Robert [31] (The Natural Step). Also, as Brendon Rowen reminds [32], influences come from the Hannover Principles and the cradle-to-cradle approach [28]. Among the Hannover Principles that are worth mentioning are the right of humanity and nature to coexist in a healthy, supportive, diverse, and sustainable condition; the recognition of interdependence; the acceptance of the responsibility for the consequences of design decisions upon human well-being; the viability of natural systems and their right to coexist; the creation of safe objects of long-term value; the elimination of the concept of waste; the idea to rely on natural energy flows; the understanding of the limitations of design; and the seek for constant improvement by the sharing of knowledge. These principles with others added later to the initial ones have been grouped into the cradle-to-cradle design protocols or standard. The C2C standard evaluates and assesses product design, processing, and manufacture criteria and administers a certification of the final product. These products are specifically designed to flow effectively through the various channels of the circular economy system.

Another system approach is the research body of academic literature that is named industrial ecology. The approach is systemic and holistic [33] so that industrial systems are considered part of and connected to higher-level systems of which have the same operational rules in terms of materials, energy, and information [34, 35]. Energy and material are flows that circulate in socioeconomic systems. Their optimum use and management [36] would derive from cultural and economic evolution as well as technological and structural changes. To this perspective, this kind of changes can derive only from innovation (production, manufacturing) and new design (products, services) developed to save resources (energy, material) and reduce the unrecyclable waste. The final goal is to close the loop, that is, to direct materials (and the embedded energy) back to production processes.

The principles of industrial ecology applied at the micro (firm) level bring in the debate and the concept of industrial symbiosis that focuses on intra-firm collaboration through market exchanges for which the waste or unused material of some firms can become resources for other firms in other industries leading to the creation of new sustainable products and markets [37, 38].

Other two bodies of knowledge organic and original to the circular economy are the product-service system theory and the blue economy. The product-service system theory states that production systems should focus on the right mix of tangible products and intangible services in order to match and satisfy the functionality or desire of the final customer. Economic and environmental positive returns match when the efforts of production are on the value created and delivered and not on the total sales [39, 40]. The second, the blue economy addresses innovation as strategic to guide businesses in developing new products and processes inspired by natural ecosystems [30].

This wide and growing body of literature has influenced over the years the concept of circular economy and the related (several) definitions. Some definitions share the idea that a circular economy emerges when closed loops of material flows and energy are created and there is evidence of reduction in the use of resources, negative environmental impacts like pollution, and waste [41] or an increase in the rates of industrial symbiosis or the quantity of reused, repaired, or transformed products. Creating closed loops systems for energy and material flows through different and innovative techniques refers, de facto, to maintain as much as possible the created value added within the economic system for obtaining the maximum value from each resource or slowing down the loss of value per item. Other definitions [25] refer to bring more attention, investments, and efforts—to make the circular economy equal or central to the sustainability transition discussion—to other fundamental aspects such as the strategic role of renewable resources, the role of energy efficiency and conservation, land management, soil protection and water, competitiveness and employment, the improvements in living and economic models, as well as social well-being.

2.2 The circular economy's concepts and definitions in society

Although already diffused and adopted especially in academia, the term circular economy was officially defined for first and then entered the wide public domain, with the establishment of the Ellen MacArthur Foundation (EMF) in 2009. The Foundation was established to focus on educating, promoting, and implementing globally the circular economy principles through different initiatives. Since then, the term has been widely spread globally.

Nowadays, the concept of circular economy refers to a new paradigm, a new way of thinking design, production, management, and consumption. In such terms, it is a relatively new system of operations whose aim is to overcome the classic and unsustainable linear production system—known as the "take-make-dispose" model.

The transition from the older model to the newer one should "close the loop" of production through the minimization of waste along the value chain. The reduction of waste along all the value chain or production system contributes first to the maintenance of value within the economic system for more time and then to the reduction in use of incineration and landfills giving waste new meaning and value for the reuse or regeneration of new products.

New and better design should lead the production system to reduce waste at each stage of the process. At the same time, the reduction and after the reuse of waste proceed together with the reduction in the depletion of resources used along the production process. The reuse of resources includes the possibility of reclamation by the original manufacturer for use in new products. In an efficient economic system, this means financial savings for the manufacturers, as well as reduction of negative ecological impacts while retaining material value.

As reported in the World Economic Forum, the MacArthur Foundation and McKinsey's report [42] as well as in the first three reports on circular economy by the MacArthur Foundation [8–10], a more specific definition refers to it as an industrial system that is restorative or regenerative by intention and design. It replaces the end-of-life concept with restoration, shifts toward the use of renewable energy, eliminates the use of toxic chemicals, which impair reuse and return to the biosphere, and aims for the elimination of waste through the superior design of materials, products, systems and business models.

The principles of this economy are:

1. To design out waste that implies the products are designed and optimized for a cycle of disassembly and reuse that is different from disposal and even recycling since there should be large losses of embedded energy and labor.

2. The strict differentiation between consumable and durable components of a product. Consumables in the circular economy refer to biological ingredients nontoxic and possibly even beneficial, which can be returned to the biosphere. Durables refer to those components made of technical nutrients such as metals and most plastics that must be designed from the start for reuse and products subject to rapid technological advance are designed for upgrade.

3. The energy required to fuel the cycle should be renewable by nature to decrease resource dependency and increase systems resilience.

These principles inform the main *four methods or ways* to reach and create value in a circular economy. These are the *power of the inner circle*, *the power of circling longer*, *the power of cascaded use*, and *the power of pure inputs*. The value should arise from the price difference between used and virgin materials (arbitrage opportunities) that show up in the production processes. The first—the power of the inner circle—refers to reaching potential savings on material, labor, energy, and capital that are embedded in the product, in the production process, and to the reduction or minimization of associated externalities like greenhouse gas (GHG) emissions, water, and toxicity. The third, the power of cascaded use, refers to diversifying reuse across the value chain of material used in the production chain. The fourth—the power of pure inputs—refers to the efficient management of uncontaminated material streams to maintain quality, extend product longevity, and increase material productivity.

3. The strategic role of the circular economy in the EU

In the final report published on March 2019 [43] on the implementation of the Circular Economy Action Plan—the EU Commission presents the main results of implementing the Action Plan adopted in 2015 and lists the future challenges the EU economy will face to reach the goal of a more competitive Europe based on a circular and climate-neutral economy where pressure on natural and freshwater resources as well as ecosystems is minimized.

The EU Commission states that the 54 actions adopted with the Circular Economy Action Plan on December 2015 have been completed. The plan was launched to realize the goal of a resource-efficient and competitive European economy. The transition to the circular economy has produced the following effects: (i) helped the EU to create more jobs (a 6% increase in 2016 in comparison to 2012 and equal to more than 4 million workers in sectors relevant to the circular economy) and (ii) opened up new business opportunities, new business models, and markets domestically and outside the EU (in 2016 circular activities—repair, reuse, and recycle—were accounted around €147 billion in value added while standing for around €17.5 billion worth of investments [44]) (The EU Monitoring Framework for the Circular Economy [45]).

The final report recognizes the strategic importance of the Action Plan for the transition to the circular economy since it promoted for the first time a systemic approach across the entire value chains and mainstreamed circular principles into different areas such as plastic production and consumption, water management, food systems, and the management of specific waste streams. Moreover, the modernization of the European industrial base—to support the EU global competitiveness—must pass through the preservation and restoration of the EU's natural capital. In other words, the circular economy is the only vision to share and invest on with the collaboration of all stakeholders, from EU institutions to member states, social organizations, and businesses.

For the benefits of the aims of this chapter, it is fundamental to contextualize the following case study showing the definitions, characteristics, and impacts the European framework on the circular economy has or should have on businesses.

3.1 The building of the circular economy and its challenges

Circular economy can be defined in different ways according to the ends of the analysis: a global trend, a challenge, an economic model, an industrial policy, new markets and innovative business models, and innovations in products and services.

The analysis of the EU documents is crucial to show the main characteristics of the circular economy that are common to all the definitions and that impact all the involved stakeholders.

First, the circular economy is a new economic system and model that requires a different way of thinking and implementing design and processes. Design is especially essential for ensuring circularity. For the EU Commission, the circular design of products is essential as it is the energy efficiency of the processes (Ecodesign Working Plan 2016–2019 [46]). One tool to guide new design investments and innovations and monitor their development and applicability is based on standardization processes and requirements such as the EU Ecodesign and energy labelling measures. These include rules on material efficiency (end-of-life treatment, ease of repair, availability of spare parts) and horizontal criteria to measure durability, reusability, reparability, recyclability, and the presence of critical raw materials. Developed by the European Standard Organization (ESO), these criteria should be applied in existing and new standards [20].

3.2 Circular design and production processes: the role of SMEs

Design is essential and promoting circular design is key to transition: new circular products and services that reduce resource use and foster materials' reuse, recovery, and recyclability are strategic to leave the old linear economic systems and models in favor of new ones. Circular products and services are produced according to different production chains and processes. The circularity aspects introduced by the Commission—energy consumption and material use, waste prevention, recycling, and reduction of hazardous chemicals—have become reference standards for the adaptation or construction on industrial plants by setting them into specific Best Available Techniques Reference documents (BREF and Industrial Emissions directive [47]). As in the case of products, industrial plants are submitted to new standards regarding circularity that become essential to obtain certification scheme as the EU EMAS (Eco-management Audit Scheme) to show organizations' environmental performance. The improvement of organizations' environmental performance is strategic to maintain, define, and increase the competitiveness of these organizations that are for the majority small- and medium-sized enterprises (SMEs). These firms are at the core of the transition. Their strategies, organizations, business models, product, and service designs are at the center of any transition to circularity. This situation asks them to invest urgently to improve their resource efficiency and production processes. Investments are not just financial or technological but first cultural: they require a change in the mindset and mentality of the entrepreneur and of the top management of firms and then of all the workers and partners of these companies.

3.2.1 Circular economy considers waste a resource

Other essential elements of a circular economy are waste management systems (WMS). The EU has invested to consolidate the European model as one of the most effective in the world (a revised waste legislative framework in July 2018 [48]) through a revised legislation that regards many themes and important topics for companies such as a clarification of legal status for recycled materials and by-products, reinforced rules and new obligations on separate collection (e.g., textiles), and the minimum requirements for the extended producer responsibility (EPR). At the core of the revised legislation, there is the challenge for the Commission to show that an efficient WMS can create and give new business opportunities. And that is especially true if firms' investments are oriented toward energy and material-efficient recovery technologies as well as policy interventions that try to make better use of economic instruments and improve planning to avoid incineration overcapacity. The efforts of the Commission have been channeled to maintain value within value chains and production processes in order to avoid unnecessary loss of valuable resources through landfilling and incineration [49].

3.2.2 Circular economy as closing loops of recovered materials

One of the objectives of the EU plan on circular economy is to increase the use of secondary raw materials (SRM). Today market operators face many challenges. The main four [50, 51] are improving substance traceability and information flows, better enforcement and use of other measures to ensure a level playing field between the EU and non-EU operators, improved harmonization and mutual recognition of end-of-waste criteria, and reinforcing circular economy aspects in instruments such as the Ecodesign directive. Knowledge and information are key to develop an efficient, transparent, and competitive market of SRM.

3.2.3 Innovation and investments and open challenges for the transition

Once the process has started, it is necessary to support it over the years, in different places, and at any level. Moreover, to reach the defined goals by the set timelines, it is fundamental to accelerate the transition. Investments in innovation and adaptation are key to make the industrial sectors ready to create more value in a competitive and environmental way. In order to support the acceleration toward transition to circular economy, the EU has disposed the Circular Economy Finance Support Platform as the main tool to manage more than €10 billion in public funding in coordination with the European Investment Bank [52].

The circular economy is now an irreversible, global mega trend the EU is investing in to make it the backbone of the EU industrial strategy to support the transition toward a circular and climate-neutral economy. The ambition to become the most competitive and sustainable region in the world requires a holistic approach and collaboration with many different stakeholders, first SMEs (strategic long-term vision for a prosperous, modern, competitive and climate-neutral economy by 2050 [53]). The EU has to manage and overcome many challenges to reach the related goals.

First, the reflection paper toward a Sustainable Europe by 2030 [54] reminds that to make the circular economy the backbone of the EU industrial policy and the EU world leader in designing and producing circular products and services:

1. Circularity should be expanded to many new areas and sectors as well as tools such as life-cycle assessments of products, the Ecodesign framework, the work on chemicals, the nontoxic environment, eco-labelling and eco-innovation, critical raw materials, and fertilizers need to be accelerated.

2. Consumers should be empowered to make informed choices.

3. Sustainable public procurement should be enhanced in the public sector.

Second, according to the 71 regions identified as at risk of missing the 2020 recycling targets or facing specific challenges, the member states and businesses must implement the revised waste legislation and develop markets for secondary raw materials (SRM) to reach the objective of maintaining the material's value within the economic system for much more time in a cost-efficient and safe way.

Third, the Action Plan states that the EU should support research, innovation, and investment in the priority sectors identified, and the holistic approach adopted for the European Strategy for Plastics in a Circular Economy should be expanded and applied to other sectors such as IT, electronics, mobility, the built environment, mining, furniture, foods and drinks, and textiles.

Fourth, the Bioeconomy Strategy [55] and the revised renewable energy framework [56] will be further steps toward using biological resources in a circular way, respecting the ecological boundaries and contributing to halting biodiversity loss.

As stated in the strategic long-term vision for a prosperous, modern, competitive, and climate-neutral economy by 2050, the transition toward a circular economy and a climate-neutral economy should be pursued together, based on a strong industrial ambition and reaping the EU businesses' first-mover advantage in these areas. Finally, climate change and energy enter the circular economy path within the efforts to cut global greenhouse gas emissions. Investment and research should lead businesses and especially SMEs to invest in new circular business models, recycling, energy and material efficiency, and new consumption patterns. The promotion of a joint approach in firms and communities to reduce production costs is key; support new partnerships between businesses, efficient use, and treatment of raw materials; and create new markets through industrial symbiosis.

In conclusion the EU has been investing to become the global leader in the circular economy through its strategy to make it the backbone of the EU industrial system and to support the transition toward a circular and climate-neutral economy. The ambition to become the most competitive and sustainable region in the world requires the collaboration of many different stakeholders, first of all businesses. The next paragraph shows the case of an artisan fashion firm investing to change its business model according to circular economy principles.

4. The case of an Italian fashion firm and its supply chain investing in circularity

The case here presented refers to a company producing quality clothing items for women. The production network is distributed in two Italian regions: in the North of Italy, in the region of Veneto, and in centre of Italy, in Tuscany between Prato and

Florence. The annual production is about 40,000 clothes. This informal network is led by an artisan micro-firm (less than ten collaborators and revenues for less than 3 million Euro). This firm produces for niche markets high-quality clothes designed and made in Italy. It was established as an innovative startup more than 10 years ago due to its innovative approach in design, communication, and use of technical performant materials (e.g., adopted in the sportswear industry). This approach was appreciated by the market and the professional world by the awards received in quality and innovation at fashion events in Italy and abroad.

4.1 The business model and the value chain structure

The value of its production resides in the original mix of design and performant material used for all the collections, not just one single garment. According to them, there are no similar companies in Italy, and that makes the company a kind of unique. Recognized international benchmarks and competitors are firms such as Maison Margiela, Ann Demeulemeester, and Yohji Yamamoto. These brands are similar for the cutting and production stages. They also use similar materials but not for all their collections. The Italian firm is more focused and specialized till the point that it is possible to say (their definition) that its designers create with an international mood but produce with an Italian artisan quality and fitting. In terms of realized volumes, the firm is more like Japanese boutiques.

The firm was established to produce quality clothing items for women in a responsible way since its beginning. The value produced and delivered by the firm relies on a researched mix of innovative design and use of selected technical material. This is made possible because the firm has a proprietary design and produces through a network of selected small and micro laboratories belonging to the typical Italian fashion districts of Veneto and Tuscany. The key resource is immaterial, the human capital or the embedded knowledge, represented by this productive system made of Italian artisans. The network is like a widespread production model, in which each subject is custodian of special knowledge and skills, which come from an ancient craftsmanship but look to the future. The value chain extends from the design of clothes made by the leading firm through the purchase of quality fabrics (usually certified synthetic material) from selected local suppliers to the widespread manufacturing system (craftsmen and specialized laboratories of the Italian territory). The chain includes partners in distribution and sales logistics (dedicated show rooms and a network of shops in Italy, it also sells abroad through e-commerce and some shops). The traditional business model has allowed this artisan firm to move from 5000 of the beginning to about 40,000 items in 2018.

4.1.1 Social innovation as responsibility: prodromes of something beyond CSR

Responsibility has characterized the firm since its beginning and has developed through the years focusing first on the social aspects of production. During the years, the social innovation aspects have been increased and delivered to the national audience through an original project for social empowerment. The project—promoting campaigns against the violence on women (e.g., genital mutilation)—has been accepted and recognized as strategic to support innovative campaigns in support of women's rights by many stakeholders. The firm has been able to create a collaborative network between the government, national NGOs and social cooperatives, national public and private media, and other fashion companies and media to promote civil messages delivered through new products made by disadvantageous people (e.g., economic or political immigrants). This experience has been received many positive feedback and awards such as participation to national conferences at the Italian Parliament or international events organized in Italy and Europe.

4.1.2 Connecting social impact and climate change concerns: how?

After a few years of experience in delivering innovative initiatives in the social field, the firm has started to think on how to expand its initial project to include climate change concerns and increase its potential impact including environmental sustainability. Circular economy seems to be the answer. From now on the company is studying how to change its business model to support its social innovation initiative through a circular economy business model: how sustaining it with a new responsible and sustainable production model based on circularity.

4.2 The circular economy project: focus on energy and material flows

The firm on the basis of external audit, analyses, and assessments has found that there are inefficiencies in the production cycle such as risks on virgin fibers (price volatility and security of supply, quality of certification schemes), low wearing time of sold clothes, and excessive quantity of unused garments after sale. More specifically, there has been ascertained the existence of unused fabric surplus, inventories of finished products, and large quantities of waste that can become resources.

The firm has developed a strategy to use resources more rationally in upstream and downstream, decoupling the consumption of materials from their use through the application of circular economy principles. The change affects the whole value chain of the company manufacturing processes. The firm intends to intervene with sectoral projects in five areas considered as strategic to transform the economic model from linear to circular and regenerative: (i) responsible supply chain, (ii) warehouse management system, (iii) circular new design, (iv) sustainable production, (v) retail 4.0. Given the complexity of the intervention, the firm has chosen to proceed in successive steps and to start from the stage relative to the management of the warehouse (flow of material) and energy management (energy flow).

The project of rationalization of the warehouse activities, which is the nodal point of production, aims to make management more efficient through the reengineering of the processes with the aid of advanced technologies and with the initiation of industrial symbiosis processes. The firm believes that in order to develop its business model in synergy with environmental protection and social progress, it is necessary to start from the management of material flows. The lower use of fabric quantities minimizes material flows in the main and inverse direction of the supply chain more than other technical solutions do, also reducing the energy flows, incorporated into products, or relating to production cycles and transportation. The quantities of waste and losses for each processing phase are also proportionally reduced. In this sense, the project envisages a proactive strategy in terms of sustainable development, acting in two directions: rationalization of the management of the raw material and a "second life" for excess quantities. The development of the project includes change in the management of the warehouse and in particular of the ordering systems (technological innovation); elaboration of guidelines for suppliers and processing cycle (social innovation); promotion of the project in the national, and international context (sustainability communication and marketing); and activation of relationships for the implementation of industrial symbiosis (economic innovation). The project aims to optimize warehouse management, reducing inventories and using the remaining materials in other industrial processes external to the company (industrial symbiosis and new product development). The optimal management of material requirements and supplies, being linked to the trend of a variable demand, requires the support of advanced technologies to reengineer the management activities. The project, therefore, activates an inter-sectoral cooperation (resource efficiency and industrial symbiosis), implements technological innovations (sustainable business model, technologies and information management, traceability of materials), and modifies the warehouse management according to the rules of circular economy (lower consumption, more efficiency). The proposed solution also allows the creation of a cross-sector network and the development of new brands for productions activated in symbiosis with other manufacturing sectors.

The expected results are reduction of production costs (warehouse costs by 10-20% in the first 2 years), expansion of markets, increase in the value of manufactured goods (increase in the ratio of revenues/units of purchased material), implementation of excellent and virtuous networks (partners in industrial symbiosis), lower risks of supplying raw materials, more renewable energy resources and higher efficiency, and more innovative design.

5. Discussion and conclusions

What is the impact of an artisan firm and its partner network investing to change its business and production model? One way to answer to this question is to discuss what has emerged from the previous analysis in terms of impact, mainly sustainable impacts.

The first impact to discuss is the *economic* one that can be read in terms of competitive advantage or market and scalability. About the opening of new markets deriving from the project, interviews to the owners make emerge the presence of some critical factors. The transition to the new business model will face: (1) speed of change; (2) acceleration dictated by the market and not by the regulatory system; and (3) availability of immediately applicable solutions (not just technological, but technical and financial). Due to its position in the firm's value chain, the project affects the business model either upstream, the supply chain (quality of raw material and related risks), or downstream, the processing cycle. In principle, the main target of the firm does not change (well educated women with high spending capacity), even if this could increase due to the introduction of circular products in the market and the activation of industrial symbiotic processes. This could allow the firm to find new partners in different sectors with whom developing new value chains. A systemic approach and a strategic network of partners become essential to the realization of positive impacts and returns. The relationships at the various levels and with the individual operators along the supply chain will continue with the traditional methods of the Italian manufacturing excellence integrated by the elaboration/evaluation of the information acquired (feedback) to create new and superior value for the "traditional" market. The type and size of the traditional market/target is presumed not to increase during the first 2 years. New markets will open up to the extent that processes of industrial symbiosis will be activated and new products will be generated in different economic sectors: there are many unknown factors at the operational level, but the firm has already identified a series of "promising" sectors for the activation of the symbiotic processes such as furniture, work clothing (technical, medical, etc.), footwear, leather goods, automotive, etc.

About the possibility to scale up solution, the project has a potential development in terms of scalability, replicability in other areas and market segments, creation of a network of "virtuous" operators in the textile sector, and growth of industrial symbiosis initiatives. The project is scalable because the parameters can be tailored to the objectives and to the typical production of the company/industry to which they apply. The project can be replicated because the conceptual plant is independent of the type of production and can be adapted to any type of product or specific production in the fashion and textile industries.

Other transversal criteria regard the *environmental and the social impacts and equal opportunities*. The environmental impact will be significant due to the reduction in consumption of nonrenewable raw materials, harmful substances released into the natural environment, and CO_2 emissions. As described before, the environmental impact coming from an efficient management system applied to the warehouse is significant for the reduction in both direct and indirect terms. In terms of social impact, the project regards hybridization of knowledge and skills; job opportunities especially in terms of new employment in favor of young employees, in cooperation with labor unions, and disadvantaged women, in cooperation with NGO; and application of SDGs criteria in the supply chain and in the production line. As it emerges from these brief lines, the project maintains central to its vision the theme of equal opportunities and access to training and knowledge since it in no way must limit participation, especially of women. On the contrary, it makes strategic the involvement of young designer artisans and women in terms of education, training, startup processes, female entrepreneurship, and social innovation.

6. Conclusions

The paradigm shift in the textile clothing sector is in a nascent state: the real competition is with the aversion to change. Implementing a circular economy model means changing policy, management, information, and finance: the transition requires a systemic and holistic approach with the involvement of many stakeholders. Cultural barriers condition the regulatory framework and produce operational difficulties. At the same time, these are conditioned by the regulatory framework itself; then, they concern the public opinion, economic operators, and final users equally. For the benefits of the project, the main operational obstacle lies in the different relationship that must be established with the supply chain and production line operators. This obstacle is overcome through a recovery of efficiency of information flows accompanied by technological innovation, the development of guidelines for suppliers, and coaching actions for the production line. The project is conceived for micro and small enterprises with the aim of enhancing their value, specificity, and excellence. It is based on circular economy principles that find a possible operational application to resource efficiency combined with technological and organizational innovation in warehouse management system. For the firm and its value chain to reach their sustainable and circular goals, it is evident that the European strategies, plans, and material helps are essential. Micro and small firms can overcome inner and contextual limits working together with strategic partners at different levels: workers unions, business association, universities, social organizations, and corporations become key for the transition of fashion micro and small enterprises to a circular and prosperous future.

Author details

Marco Tortora[1,2,3]* and Giuseppe Tortora[2]

1 Polimoda International Institute of Fashion Design and Marketing (Polimoda), Florence, Italy

2 FAIR Association, Florence, Italy

3 University of Florence, Florence, Italy

*Address all correspondence to: marco.tortora@gmail.com

IntechOpen

© 2020 The Author(s). Licensee IntechOpen. Distributed under the terms of the Creative Commons Attribution - NonCommercial 4.0 License (https://creativecommons.org/licenses/by-nc/4.0/), which permits use, distribution and reproduction for non-commercial purposes, provided the original is properly cited. CC BY-NC

References

[1] European Commission. Fashion and High-End Industries in the EU. 2015. Available from: http://ec.europa.eu/growth/sectors/fashion/textiles-clothing/eu/index_en.htm [Retrieved: January 2019]

[2] European Commission. Textiles and Clothing in the EU. 2015. Available from: http://ec.europa.eu/growth/sectors/fashion/textiles-clothing/eu/index_en.htm [Retrieved: January 2019]

[3] Euroactiv. European Textiles and Fashion: Facts & Figures. 2019. Available from: https://www.euractiv.com/section/innovation-industry/infographic/european-textiles-and-fashion-facts-figures/ [Retrieved: March 2019]

[4] Statista. Global Apparel Market Size Projections from 2012 to 2025, by Region (in Billion U.S. Dollars). 2018. Available from: https://www.statista.com/statistics/279757/apparel-market-size-projections-by-region/ [Retrieved: January 2019]

[5] Remy N, Speelman E, Swartz S. Style that's Sustainable: A New Fast-Fashion Formula. 2016. Available from: https://www.mckinsey.com/business-functions/sustainability-and-resource-productivity/our-insights/style-thats-sustainable-a-new-fast-fashion-formula?cid1/4sustainability-eml-alt-mip-mck-oth-1610

[6] Ellen MacArthur Foundation. A New Textiles Economy: Redesigning Fashion's Future. Cowes, Isle of Wight: Ellen MacArthur Foundation; 2017. Available from: https://www.ellenmacarthurfoundation.org/publications/a-new-textiles-economy-redesigning-fashions-future [Retrieved: January 2019]

[7] Ellen MacArthur Foundation. The New Plastics Economy: Rethinking the Future of Plastics. Cowes, Isle of Wight: Ellen MacArthur Foundation; 2016

[8] Ellen MacArthur Foundation. Towards the Circular Economy 1: Economic and Business Rationale for an Accelerated Transition. Cowes, Isle of Wight: Ellen MacArthur Foundation; 2012 [Retrieved: January 2012]

[9] Ellen MacArthur Foundation. Towards the Circular Economy 2: Opportunities for the Consumer Goods Sector. Cowes, Isle of Wight: Ellen MacArthur Foundation; 2012 [Retrieved: January 2013]

[10] Ellen MacArthur Foundation. Towards the Circular Economy 3: Accelerating the Scale-up Across Global Supply Chain. Cowes, Isle of Wight: Ellen MacArthur Foundation; 2012 [Retrieved: January 2014]

[11] McKinsey. Europe's Circular Economy Opportunity. 2015. Available from: https://www.mckinsey.com/business-functions/sustainability-and-resource-productivity/our-insights/europes-circular-economy-opportunity [Retrieved: January 2019]

[12] Wijkman A, Skanberg K. The Circular Economy and Benefits for Society. The Club of Rome; 2016

[13] European Commission. Circular Economy in Practice—Reducing Textile Waste. 2017. Available from: https://ec.europa.eu/easme/en/news/circular-economy-practice-reducing-textile-waste [Retrieved: January 2019]

[14] Leal Filho W, Ellams D, Han S, Tyler D, Boiten V, Paco A, et al. A review of the socio-economic advantages of textile recycling. Journal of Cleaner Production. 2019

[15] European Commission. Closing the Loop—An EU Action Plan for the

Circular Economy. Communication from the Commission to the European Parliament, the Council, the European Economic and Social Committee and the Committee of the Regions; COM (2015) 614 final; 2015

[16] European Commission. Commission Staff Working Document: Additional Analysis to Complement the Impact Assessment SWD (2014) 208 Supporting the Review of EU Waste Management Targets; 2015

[17] European Commission. Communication on the Review of the List of Critical Raw Materials for the EU and the Implementation of the Raw Materials Initiative. Communication from the Commission to the European Parliament, the Council, the European Economic and Social Committee and the Committee of the Regions; COM (2014) 297 final; 2014

[18] European Commission. Report on the Implementation of the Circular Economy Action Plan. Brussels; COM (2019) 190 final; 2019. Available from: https://eur-lex.europa.eu/legal-content/EN/TXT/?qid=1551871195772&uri=CELEX:52019DC0190 [Retrieved: 4 March 2019]

[19] European Commission. Commission Staff Working Document. Accompanying the Document. Report on the Implementation of the Circular Economy Action Plan. Brussels; SWD (2019) 190 final; 2019. Available from: https://eur-lex.europa.eu/legal-content/EN/TXT/HTML/?uri=CELEX:52019SC0090&from=EN [Retrieved: 4 March 2019]

[20] European Commission. Commission Staff Working Document. Sustainable Products in a Circular Economy—Towards an EU Product Policy Framework Contributing to the Circular Economy. Brussels; SWD (2019) 91 final; 2019. Available from: https://eur-lex.europa.eu/legal-content/EN/TXT/HTML/?uri=CELEX:52019SC0090&from=EN [Retrieved: 4 March 2019]

[21] European Commission. Circular Economy. Closing the Loop [Press Release]. 2019. Available from: https://ec.europa.eu/commission/sites/beta-political/files/circular-economy-factsheet-general_en.pdf [Retrieved: May 2019]

[22] Boulding K. The economics of the coming spaceship earth. In: Jarrett H, editor. Environmental Quality in a Growing Economy. Baltimore, MD: Johns Hopkins University; 1966

[23] Pearce DW, Turner RK. Economics of Natural Resources and the Environment. Hemel Hempstead: Harvester Wheatsheaf; 1990

[24] Lieder M, Rashid A. Towards circular economy implementation: A comprehensive review in context of manufacturing industry. Journal of Cleaner Production. 2016;**115**:36-51

[25] Rizos V, Rizos V, Tuokko K, Tuokko K, Behrens A, Behrens A. The circular economy: A review of definitions, processes and impacts. CEPS Research Report No 2017/8; 2017

[26] Braungart M, McDonough W, Bollinger A. Cradle-to-cradle design: Creating healthy emissions—A strategy for eco-effective product and system design. Journal of Cleaner Production. 2006;**15**:1337-1348

[27] Ankrah NA, Manu E, Booth C. Cradle to cradle implementation in business sites and the perspectives of tenant stakeholders. Energy Procedia. 2015;**83**:31-40

[28] McDonough W, Braungart M. Cradle to Cradle: Remaking the Way We Make Things. New York, NY: North Point Press; 2002

[29] Stahel WR. The Performance Economy. NY: Palgrave; 2006

[30] Pauli G. The Blue Economy—10 Years, 100 Innovations, 100 Million Jobs. Taos, NM: Paradigm Publications; 2010

[31] Robèrt KH, Anderson R. The Natural Step Story: Seeding a Quiet Revolution; 2002

[32] Rowen B. Is UK doing enough on the circular economy? [Online Article]. 2019. Available from: https://www.governmenteuropa.eu/is-the-uk-doing-enough-on-the-circular-economy/93304/ [Retrieved: March 2019]

[33] Garner A, Keoleian GA. Industrial Ecology: An Introduction. Ann Arbor, MI: National Pollution Prevention Center for Higher Education; 1995

[34] Erkman S. Industrial ecology: An historical view. Journal of Cleaner Production. 1997;**5**(1-2):1-10

[35] Ehrenfeld JR. Would industrial ecology exist without sustainability in the background? Journal of Industrial Ecology. 2007;**11**(1)

[36] Graedel TE, Allenby BR. Industrial Ecology. Englewood Cliffs, NJ: Prentice Hall; 1995

[37] Chertow MR. Industrial Symbiosis: Literature and taxonomy. Annual Review of Energy and the Environment. 2000;**25**:313-337

[38] Lombardi RD, Laybourn P. Redefining industrial symbiosis. Crossing academic-practitioner boundaries. Journal of Industrial Ecology. 2012;**16**(1):28-37

[39] Tukker A. Product services for a resource-efficient and circular economy—A review. Journal of Cleaner Production. 2015;**97**:76-91

[40] Tukker A, Tischner U. Product-services as a research field: Past, present and future: Reflections from a decade of research. Journal of Cleaner Production. 2006;**14**:1552-1556

[41] Sauvé S, Bernard S, Sloan P. Environmental sciences, sustainable development and circular economy: Alternative concepts for trans-disciplinary research. Environmental Development. 2016;**17**:48-56

[42] World Economic Forum. Towards the Circular Economy: Accelerating the Scale-Up across Global Supply Chains. Geneva, Switzerland; 2014

[43] European Commission. Report from the Commission to the EU Parliament, the Council, the EU Economic and Social Committee of the Regions on the Implementation of the Circular Economy Action Plan. Brussels; COM (2019) 190 final; 2019 [Retrieved: 4 March 2019]

[44] Eurostat. 2016. Available from: https://ec.europa.eu/eurostat/tgm/refreshTableAction.do?tab=table&plugin=1&pcode=cei_cie010&language=en [Retrieved: January 2019]

[45] European Commission. Report from the Commission to the EU Parliament, the Council, the EU Economic and Social Committee of the Regions on a Monitoring Framework for the Circular Economy. Strasbourg; COM (2018) 29 final; 2018 [Retrieved: 16 January 2018]

[46] European Commission. Communication from the Commission Ecodesign Working Plan 2016-2019. Brussels; COM (2016) 773 final; 2016 [Retrieved: 30 November 2016]

[47] European Parliament. Directive 2010/75/EU of the European Parliament and of the Council on Industrial Emissions (Integrated Pollution Prevention and Control); 2010

[48] European Commission. Directive 2008/98/EC on Waste; EU Directive 1999/31/EC on the Landfill of Waste; Directive 94/62/EC on Packaging and Packaging Waste; Directive 2000/53/EC on End-of Life Vehicles; Directive 2006/66/EC on Batteries and Accumulators and Waste Batteries and Accumulators; Directive 2012/19/EU on Waste Electrical and Electronic Equipment (WEEE); 2008

[49] European Commission. Communication from the Commission to the European Parliament, the Council, the European Economic and Social Committee and the Committee of the Regions. The Role of Waste-To-Energy in the Circular Economy. Brussels; COM (2017) 34 final; 2017 [Retrieved: 26 January 2017]

[50] European Commission. Communication from the Commission to the European Parliament, the European Council, the Council, the European Economic and Social Committee and the Committee of the Regions. The Single Market in a Changing World. A Unique Asset in Need of Renewed Political Commitment. Brussels; COM (2018) 772 final; 2018 [Retrieved: 22 November 2018]

[51] European Commission. Communication from the Commission to the European Parliament, the Council, the European Economic and Social Committee and the Committee of the Regions. Implementation of the Circular Economy Package: Options to Address the Interface between Chemical, Product and Waste Legislation. (Text with EEA Relevance). Strasbourg; COM (2018) 32 final; 2018 [Retrieved: 16 January 2018]

[52] European Commission. Communication from the Commission to the European Parliament, the European Council, the Council, the European Central Bank, the European Economic and Social Committee and the Committee of the Regions. Action Plan: Financing Sustainable Growth. Brussels; COM (2018) 97 final; 2018 [Retrieved: 8 March 2018]

[53] European Commission. Communication from the Commission to the European Parliament, the European Council, the Council, the European Economic and Social Committee, the Committee of the Regions and the European Investment Bank. A Clean Planet for All. A European Strategic Long-Term Vision for a Prosperous, Modern, Competitive and Climate Neutral Economy. Brussels; COM (2018) 773 final; 2018 [Retrieved: 28 November 2018]

[54] European Commission. Reflection Paper towards a Sustainable Europe by 2030. Brussels; COM (2019) 22 final; 2019 [Retrieved: 30 January 2019]

[55] European Commission. Communication from the Commission to the European Parliament, the Council, the European Economic and Social Committee and the Committee of the Regions. A Sustainable Bioeconomy for Europe: Strengthening the Connection between Economy, Society and the Environment. Brussels: COM (2018) 673 final; 2018 [Retrieved: 11 October 2018]

[56] European Parliament. Directive 2018/2001 on the Promotion of the Use of Energy from Renewable Sources; 2018

Chapter 6

Harnessing Small Country Collaboration Opportunities to Advance Energy Innovation and Joint Investments

Anneliese Gegenheimer and Charles Michael Gegenheimer

Abstract

Greater international collaboration is required to catalyze research and development (R&D) investment flows in energy technologies. Successful deployment of such technologies requires innovative funding mechanisms, intellectual property, and data-driven analyses to make smarter, sustainable investment decisions. As small countries are increasingly dealing with effects of climate change, some are projected to lose large portions of their economy. This chapter discusses ways that smaller countries, both in the developed and developing world, can harness international cooperation to advance energy innovation and mitigate such impact. In contrast to collaboration with larger countries, smaller country collaboration can build more agile, balanced partnerships in which participating countries co-develop and co-own R&D and training, and define pilot programs that target their own needs. Leveraging each other's strengths, small countries can become catalysts for global change. Smaller country collaboration is explored through a proposed model of collaboration in energy innovation between Singapore and Estonia, often considered gateways to Southeast Asia and the EU plus Russia, respectively. Specifically, Singapore and Estonia have the opportunity to leverage each other's startup ecosystems, innovation systems, knowledge-based economies, and regional markets to build a niche in clean energy technologies, particularly energy storage innovation, with potential global impact on larger markets.

Keywords: energy innovation, technological advancement, small country collaboration, bilateral partnership, Singapore, Estonia

1. Introduction

Greater international coordination and collaboration is required to research, develop and deploy reliable and affordable energy technologies critical to achieving global emission reduction targets. In turn, successful deployment of such technologies may require innovative funding, intellectual property, and data sharing strategies [1]. As small countries are increasingly dealing with effects of climate change, some are projected to lose large portions of their economy to climate change, which will require creative solutions to help mitigate this impact [2, 3]. Innovative clean energy technologies can enable countries to achieve their carbon emission reduction goals and empower them to set more ambitious goals for the future [4]. For small countries, one opportunity cost of not implementing innovative clean energy technologies is the impact on projected carbon emission reduction goals.

Fewer self-funded opportunities for independent innovation and research within smaller countries provide an incentive for these countries to seek opportunities for international joint innovation and collaboration [5]. While small countries often collaborate with larger countries that have more research and development (R&D) funding, technical assistance, and expertise to bring new technologies to market, such cross-border collaboration can lead to imbalances in ownership, accountability, and decision making.

This chapter introduces the concept of small country-to-county joint innovation, discusses the benefits and criteria for collaboration and concludes with a proposed example of joint partnership in energy storage innovation between Singapore and Estonia. Smaller country collaboration can build more agile, balanced partnerships in which participating countries co-develop and co-own R&D and training, and define pilot programs that target their own needs. Leveraging each other's strengths, small countries can become catalysts for global change.

2. Why small countries matter

According to French writer Milan Kundera, "What distinguishes the small nations from the large is not the quantitative criterion of the number of their inhabitants; it is something deeper. For the small nations, existence is not a self-evident certainty but always a question, a wager, a risk; they are on the defensive against history, that force which is bigger than they, which does not take them into account, which does not even notice them [6]."

Kundera's assertion is backed up by the creators of the Global Innovation Index, which measures a country's innovation impact based on innovation inputs (such as higher education) and outputs (such as knowledge creation) [7]. As the world has increasingly globalized, small countries such as Singapore, Switzerland, and the Baltic countries have become "innovative by necessity [8]." Larger countries, on the other hand, innovate out of ambition and competition. This common attitude towards innovation and survival among small countries creates shared goals through which strong international partnerships can grow. Even large countries and regions such as the United States, China and the European Union stand to benefit from such partnerships and cooperation if their firms can design better products for fast-growing small country economies.

Joint innovation partnerships between small countries also require a form of matchmaking that differs from collaboration with larger countries. First, the countries must understand the comparative advantages and national innovation systems of the other to determine whether, between the two countries, they have the expertise and resources to co-innovate on any given project. Second, small country collaboration will presumably require each to risk a larger proportion of their own resources than if they were collaborating with a larger country. However, each country may enjoy greater benefits from success. This section describes the benefits of small country joint innovation and the criteria for creating a balanced partnership.

2.1 Why small countries should pursue joint energy innovation

Climate change is increasingly affecting all countries with varying degrees of impact, and current climate models point to carbon dioxide emissions as a primary factor causing climate change. Individually, small countries, both developed and developing, are helping to lead the charge. For example, while Switzerland's overall emissions are small relative to larger countries, its emissions per capita are above average. Thus, the country believes it has a responsibility to reduce its carbon emissions, and influence other countries, such as Brazil, to do the same [9]. In Eswatini, a low carbon emitting country, the country showed leadership and innovation in reducing hydrochlorofluorocarbons (HCFCs). HCFCs are a type of greenhouse gas used in refrigeration and air conditioning. The country worked in partnership with the United Nations, a German environmental organization, and Eswatini's local refrigerator manufacturing facility to phase out the use of HCFCs [10].

While the examples above do not represent small country joint collaboration, they do demonstrate the will of small countries to take part in the call to action against climate change, which has the potential to lead to future small country collaboration. Collaborative joint energy innovation and investments between small countries has the potential to have impact on:

- National policy: carbon dioxide reductions from clean energy technology implementation can potentially impact a greater proportion of a small country's energy resources and advance its energy goals.

- External support: small countries that are nimble, early adopters of new technologies can attract commercially-focused suppliers eager to assist successful demonstrations of innovative products.

- Investment focus: small countries can maximize the value of their joint investments when serving as test beds for innovative technologies focused on their own needs.

- Economic growth: the combination of smaller countries' capabilities and resources in joint projects can lead to economic growth in sectors where each has existing strengths, with mutual benefit from successful technology propagation.

- Future success: successful collaboration in one area, such as energy innovation, can lead to other opportunities for smaller countries, their industries and their people to further collaborate in other areas with mutual benefit.

These reasons are especially attractive for small countries, such as small island developing states (SIDS) that are most vulnerable to the effects of climate change. For SIDS, such as Tuvulu, that are at risk of disappearing from sea level rise due to climate change, joint collaboration in energy innovation could be critical to its survival. Successes in joint innovation between small countries that are dealing with similar challenges can be applied to other small countries as well as larger countries or regions dealing with the same challenges [11].

2.2 Challenges that small countries face

Small countries face several challenges that joint collaboration with larger countries or multilateral organizations often address. Many are finding innovative ways to address the majority of these issues, which could be used in small country joint collaborations to fill the gap left by the traditional role of larger countries. These challenges include:

1. Scale: smaller domestic markets typically limit economies of scale, which makes scaling up and deploying a new innovation more expensive. Larger countries remedy this issue by bringing access to large markets and deployment opportunities. However, some small innovative countries are often gateways to regional markets, which, with the right partnership and connections, can be a catalyst for scaling innovation. In the clean energy sector, according to the World Resources Institute, no one country dominates, signaling that there is still opportunity for small countries to play a role in regional or global supply chains [12].

2. Lack of global influence: larger countries tend to have more influence in international institutions and decision-making, making it harder for smaller countries to be heard on a global stage. In addition, small and developing countries run the risk of playing a "marginal and subordinate role" in international collaboration networks, making it difficult for them to influence the research agenda [13]. Increasingly, institutions such as the United Nations have started holding panels on small country success stories [14]. This type of international platform for small countries could incentivize more small countries to develop successful, innovative models for joint collaboration.

3. Lack of administrative capacity: according to a study on small states, innovation, and administrative capacity, small countries are particularly challenged by the administrative needs of growing international networks and influence that are a by-product of innovation [15]. One way that Israel deals with this challenge is that it drives much of its research in collaboration with research laboratories in other countries to free up its own R&D funding and capacity for other parts of its innovation system [16].

4. Lack of technological expertise or human capital: larger, developed countries have a larger talent pool for R&D. While small countries typically cannot compete on sheer quantity of talent, several small countries have programs that leverage both domestic and foreign talent. For example, Singapore has a funding program for nationals that incentivize them to return to the country after pursuing graduate studies in foreign countries [17]. Estonia takes a different approach, and encourages its entrepreneurs to start or join startups abroad, which could ultimately bring business and economic growth back to the country [17].

5. Lack of financial capacity to fund R&D: for smaller countries, limited domestic financial resources and harder-to-access international financing can stifle energy innovation. Several innovative financing options are discussed in Section 3.1.

6. Rapid changes in innovation: depending on their structure, small country economies could be severely impacted by 'creative destruction' caused by the cycle of innovation and technological change in which new jobs, industries, and products replace old ones [15]. Larger countries can afford to be more resilient to these types of changes, which makes 'creative destruction' less risky financially and with regard to skills and products. Small countries can mitigate the effects of 'creative destruction' by establishing strong yet flexible institutional capacity that can adapt to changes in innovation and the resulting demand for new skills and products [17]. In addition, a stronger focus on downstream innovation, which focuses on commercialization of early-stage

innovation, will help mitigate the risk of an innovation succumbing to the 'valley of death'. This focus could slow down the pace of 'destruction' by helping to prolong and diversify resources across the innovation cycle.

7. Overspecialization: limited domestic markets, fewer resources, and dependence on exports creates a risk for small countries to overspecialize in a certain product or skill, making the country less resilient to changes in innovation and less diverse economically [15]. Although small countries often use specialization to drive economic growth, they can ensure that policy incentives and funding mechanisms incentivize innovation to move up the value chain to avoid lock in.

2.3 Benefits of small country cross-border innovation collaboration

Small country collaboration can provide several benefits to participating countries that they would not be able to have with larger country partnerships.

First, while the proportion of resources, and thus risk is higher with small country-to-country innovation, small countries can co-innovate and deploy technologies with bigger impact to a network. In larger countries, pilots are often proven on a small-scale in communities or parts of communities. In some cases, for example, pilot programs do not see the same impacts and same benefits at scale, partially because of regulatory, economic, or constraints that were not imposed on the pilot [18]. In smaller countries, while pilots might be at a smaller or similar scale overall compared to a larger country, a deployment can demonstrate impact on a full national or sub-national ecosystem, including government, economy, and community networks, rather than a more siloed pilot program.

Second, small country collaboration can be more nimble and efficient. Similar to the difference between a large corporation and a small business, smaller countries have less 'red tape' to navigate, and can make decisions and deploy resources more quickly, leading to more a more flexible and effective partnership. More nimble partnerships could in turn facilitate more frequent information sharing and greater formal and informal networks for knowledge transfer. This, in turn, can lead to greater communication and an opportunity for the countries to help each other improve their innovation systems. However, for example, in imbalanced partnerships between developed and developing countries, knowledge transfer often does not happen due to a more passive role by the developing country [19].

Third, smaller countries can co-develop and co-produce research, training, and pilots that will truly benefit their needs, and allow them to be empowered through decision making, leadership, ownership, and financial accountability. These countries can build more targeted partnerships that will truly leverage the other's strengths and build impact and value for each other through development of technical skills and technology transfer capability. One research study observed that collaborations with small scientific communities tend to be restricted to specific fields of research that are directly linked in some way to participating countries [19]. For example, small countries use specialization in R&D intensive sectors, such as energy and technology, to drive economic growth [15]. However, these types of opportunities allow small countries to find direct or complementary linkages that are relevant to a larger audience. These linkages create opportunity for small countries to join forces, engage together with larger countries or markets, and bring scale to their innovation. For example, as we discuss in Section 4, Estonia's innovation in cybersecurity and Singapore's electric grid testbed creates opportunities for joint innovation for grid security.

Fourth, small country joint collaboration can open up opportunities to learn lessons and share successes from each other, specifically that address challenges unique to small countries. For example, small countries may not have to worry about how its R&D decisions will influence global trade flows, but they may want to share ideas about how to influence decision making in international organizations that are dominated by large countries.

Fifth, and finally, according to a joint United Nations Intergovernmental Panel on Climate Change (IPCC) and ClimateXChange panel on climate change action in small countries, small country joint collaboration can spur global advocacy and activism leadership on environmental issues [14]. For example, Scotland has leveraged international cooperation to inspire other small countries to reduce their greenhouse gas emissions through innovative initiatives [14]. SIDS countries have also joined together to draw attention to the need for innovation in addressing climate impacts. Together, small countries can make their voices heard on an international stage.

2.4 Criteria for creating a balanced partnership

Traditional cross-border innovation collaboration often focuses on bilateral and multilateral partnerships among international organizations' member countries, where collaboration arises between developed and developing countries, and regional cooperation. While small country-to-country collaborations already have a place in the sphere of cross-border joint innovation, opportunities exist outside the traditional paradigms of collaboration. Smaller countries may collaborate to apply their respective capabilities to new technology developments of high relevance and importance to them. For example, a study on Norway's innovation system found that international collaboration between foreign and Norwegian researchers on scientific research increased from 23% to 53% between 1985 and 2004 [20].

However, this trend reflects that international collaboration in general, as measured by co-authorship and publications, has been growing over time, with co-authored articles doubling over the past two decades [19]. While reasons for this trend range from increasing globalization to the rise of the internet, small countries typically rank highly in international collaboration based on this measure. Research shows an inverse relationship between the size of a country and international collaboration [5].

While this measure of co-authorship and publication is a verifiable and measurable method of determining international collaboration, it does not measure the level of contribution by each participating country, nor the level of joint innovation that truly occurs. A study on the effect of unbalanced international collaboration on a country's real contribution to scientific output found that countries with smaller scientific communities produced an 'insignificant' level of scientific output in unbalanced partnerships [19].

International collaboration, especially in the form of joint innovation, is an especially valuable opportunity to actively contribute to solving world challenges, such as food security, water issues, and energy security. Just as it's important to have inclusive participation and decision making in all levels of government, for example, small countries too should have a seat at the table. Smaller country collaborations are more likely to produce balanced partnerships.

Some research suggests that the topic of collaboration has some bearing on the type of partnership. A study of 20 countries of varied size found that collaboration on 3D printing technology yields more balanced collaboration, big data technology shows a more radial pattern with the United States in the center, and carbon nanotubes and graphene technology indicates "small-world," clustered networks [21].

While none of these technologies are specifically energy technology, researchers should be aware in advance of how their target area of research might yield preferences towards a certain type of partnership.

In addition, the research supported their hypotheses that the bigger the country, as measured by their level of knowledge reserves and the level of R&D full time equivalent (FTE) innovators, the less incentivized the country are to engage in international joint collaboration [21].

What criteria are important to building a successful small country-to-country collaboration partnership? Not every small country partnership combination is ideal–just like any partnership, it depends on the goals and commitment of each country. Joint innovation and collaboration requires a level of resourcing and commitment that not every country may have.

Often, geographical location, known as the 'neighborhood effect', can spark direct, indirect, and spillover effects between small countries neighboring each other [22]. While not a requirement, geographic proximity can come with built in mutual benefits (cross-cultural understanding, similar time zones, etc.).

Regardless, small countries with certain characteristics may have higher chances of successful cross-border partnerships.

These characteristics include [17, 21]:

- Openness to outside ideas and opportunities
- Strong and flexible institutions
- Regulatory and policy environment that supports the end-to-end innovation cycle
- Strong talent pool with engaged research network

The Institute for Management Development (IMD) publishes a world competitiveness yearbook every year, which ranks countries based on their investment and development, appeal and readiness [23]. This ranking includes investment in education, quality of life, and opportunities for career advancement. In IMD's 2019 ranking, several small countries were in the top 10, including Singapore, Qatar, and Switzerland [23]. Other rankings, such as the World Intellectual Property Organization's Global Innovation Index, World Economic Forum's Enabling Trade Index, and other indicators of innovation and entrepreneurship can help identify small countries that have criteria for a successful cross-border partnership.

3. Key strategies for successful joint innovation

Successful small country joint innovation will require purposeful shifts from traditional approaches to development, including embracing open innovation, enabling a strong investment environment, fostering an entrepreneurial ecosystem domestically, and engaging with collaborators under win-win intellectual property strategies. These shifts will enable the flexibility and equal voice that small countries need in joint collaboration, and may also benefit these countries economically well beyond the scope of energy technology innovation. This section discusses key enabling strategies for successful small country joint collaboration, which includes innovative funding mechanisms, intellectual property tools, and creative data sharing.

3.1 Funding mechanisms

Developed countries have a history of funding partnerships between developed and developing countries, which often start with the developed country first developing a study and then engaging with developing country partners for pilots or testing grounds. Traditionally, funding comes from government agencies, research councils, or other sources with R&D-focused goals. But these programs can end when the funding ends, absent committed resources in developing countries; or can result in loss of control to large countries that retain critical skill sets and establish outside economic or political influence.

Small country-to-small country energy innovation requires funding and resource commitments that become a priority for the countries and lead to sustainable initiatives. Innovative funding approaches to funding can include:

- Leveraging regional funding initiatives, such as Horizon 2020 or Mission Innovation, which coordinates global RD&D for clean energy [4].

- Jointly funding an innovation challenge that would create a pull incentive for innovators to propose new ideas.

- Directly funding collaborative domestic capabilities that support the research goals, thereby shifting the accountability to agencies and individuals on both sides that are motivated to successfully develop a more collaborative and balanced partnership [24].

- Developing Public-Private Partnerships (PPPs) that provide attractive investment of private funding to build sustainable infrastructure.

- Engaging in international joint energy technology innovation partnerships that attract venture capital funding. For example, SkeletonTech, an Estonian supercapacitor startup, originally received funding from an Estonian-Norwegian joint energy technology innovation program in 2013 and has received at least one round of venture capital funding [25].

- Focusing on downstream innovation, such as the deployment and scale-up of successful innovations through business development organizations. Business development organizations can create business plans that clarify the supply chain, value chain, path to market, and regulatory hurdles and help assure timely access to target markets for energy innovation to avoid commercialization "valley of death."

- Establishing entrepreneurial incubators and special grants for piloting promising innovative technologies or necessary building blocks to avoid the technology development "valley of death."

Each of these mechanisms can help enable small country joint innovation and provide opportunities for small countries to find funding that will make joint innovation possible without larger countries or multilateral institutions.

3.2 Intellectual property

Innovative approaches in intellectual property (IP) that foster investment in new technologies will facilitate intra-country commitments, technology transfer, and private in-country investment needed to introduce and establish a sustainable technology innovation.

Patenting and licensing should be at a pace consistent with commercial opportunities, and should set forth a clear plan for rapid technology transfer that benefits both countries. The commercial upside for both countries will require standards for protection against loss of technical information and limit any unwanted diffusion of energy technologies through industries and continents [1].

One licensing example is the cluster approach, in which IP that is developed through publicly-funded projects is made available through commercial licensing to other organizations [1]. In addition, PPPs should establish up-front expectations of efficient and rapid technology diffusion so that private sector concerns with risk mitigation and IP protection are proactively addressed to protect private investment and do not hinder the diffusion process.

IP rights ownership should be planned up front for a win-win model, and while this includes IP ownership by partners who generate new innovations, the plan needs to assure the costs of IP protection, IP rights and economic returns are shared to facilitate collaboration.

Where IP rights are jointly owned by several countries or partners whose contributions are interdependent, as would be expected in a small country-to-small country collaboration and related PPPs, each country and participant would be expected to grant non-exclusive, royalty-free rights necessary to enable small countries to implement innovations in-country, and to provide a commercially reasonable sharing of economic returns from government and private investment in new technology deployed outside their respective countries.

The goal of such win-win collaboration in innovation is to enable each partner to obtain rights to use the project results that maximize commercial exploitation and share in commercially reasonable returns subject to additional agreements. One example from large country collaboration is the US–China Clean Energy Research Center, which aims "to accelerate the pace of innovation in clean energy technologies [1]." The organization has had success through clear guidelines for licensing, joint ownership, and dispute resolution.

3.3 Data sharing

Data sharing, enabled through open and flexible data flows between countries, can be a critical component in all stages of innovation. For example, in marine energy projects funded by the U.S. Department of Energy, international researchers help determine how data will be shared on a global level [26]. Since small countries do not benefit from the volume of data and number of users that larger countries have, data sharing or data-focused trade agreements can help incentivize small country joint innovation opportunities [26]. In addition, data sharing and joint collaboration can help with developing metrics to evaluate the value and potential of energy technologies in the R&D phase. Finally, as innovation in energy technology reaches the commercialization stages, data sharing can help inform standards development for new technologies [1].

4. Case study: Singapore-Estonia joint energy storage innovation collaboration

The model of smaller country collaboration is explored through proposed energy innovation collaboration between Singapore and Estonia. Singapore and Estonia are often considered gateways to markets in Southeast Asia and the EU plus Russia, respectively. Specifically, Singapore and Estonia have the opportunity to collaborate and leverage each other's startup ecosystems, innovation systems, knowledge-based economies, and regional markets to build a niche in energy storage innovation with potential global impact on larger markets [27].

4.1 The Singapore-Estonia connection

This section outlines Singapore and Estonia's current international cooperation, which reveals regulatory and business environments that can welcome collaboration on energy storage technologies. Currently, there is no explicit energy technology innovation cooperation between Singapore and Estonia [28]. In the past, Estonia has looked at the regulatory sandbox approach adopted by the Energy Market Authority (EMA) in Singapore, but found that the current regulatory framework in Estonia is flexible enough for starting businesses and there is no explicit need for a regulatory 'safe space' such as the one in Singapore [28]. Singapore's regulatory sandbox model was developed in 2018 to support energy innovation in generation, transmission and distribution, and creates a trial environment in which third parties can test energy solutions without being subject to regulatory requirements [29]. This sandbox allows promising innovations that may not comply with current regulatory requirements to be tested and deployed and allows the EMA to assess potential impacts of new technologies when deciding whether to modify or add new regulations.

However, in other areas, such as entrepreneurship, the two countries have found growing opportunities for collaboration. Enterprise Estonia, which negotiates and manages investment from outside of Estonia to its country with counseling programs for startups, opened its Singapore office in September 2016 [30]. The office promotes trade relations with Singapore and connects the booming startup community in Estonia with Asian venture capital, and uses Singapore as a launch pad to the South-East Asian (SEA) region for Estonian startups. Estonia in return offers opportunities for e-Residency to SEA business people for easy access to the EU and for managing their EU businesses [31].

In 2016, Estonia simplified its e-residency program, which allows Singaporean e-resident applicants to pick up their cards directly in Singapore instead of at an Estonian embassy in another country as Estonia does not have an embassy in Singapore. The e-residency allows entrepreneurs to establish and operate a company in Estonia remotely, and is the most efficient way of getting benefits like easy access to the EU market, e-banking services, and a streamlined digital administrative system. Thus, a Singaporean entrepreneur can establish an Estonian company that he runs from Singapore, to serve clients based across the European Union [30]. According to Estonia's chief information officer (CIO) Taavi Kotka, Singapore is one of Estonia's highest priorities in terms of collaborating with developers and service providers in one of the top global startup ecosystems in the world [32].

In addition, in January 2018, Estonia and Singapore signed an agreement on cooperation between the countries by which the countries create the possibility for joint exercises at the cyber practice fields in both countries [33]. In May, the Estonian Defense Forces Cyber Range provided a cybersecurity training at Singapore's Cyber Defense Test and Evaluation Centre (CyTEC) [33]. While no explicit energy technology innovation collaboration is currently planned between the two countries, the two countries have developed a relationship that would welcome energy technology innovation collaboration.

4.2 The case for energy storage innovation collaboration

This section makes the case for joint collaboration specifically focused on innovation in energy storage between Estonia and Singapore. The three sub-sections focus on domestic energy storage markets, regional energy storage markets, and how Estonia and Singapore could benefit from joint innovation in this sector.

4.2.1 Domestic markets for energy storage

Although Estonia's government lacks a near-term goal explicitly focused on energy storage, several startups and universities in the country are bringing energy storage technologies through the innovation pipeline, from R&D to commercialization. The country uses its ecosystem of universities, venture capital, startup culture, and science and technology park, Tehnopol, to spur innovation and primarily develop products for export [34]. Estonia's climate roadmap, however, suggests that energy storage innovation may be key for the country's future. Estonia is currently the most energy independent country in the EU due to its abundance in oil shale, but has the highest energy intensity of all OECD countries [35].

In order to reduce its reliance on oil shale, which is responsible for 90% of electricity generation and 80% of Estonia's greenhouse gas emissions, the country is looking to diversify via investment in renewables and will heavily rely on wind power by 2040 [36]. By 2025, a significant portion of shale power generation units in Estonia are scheduled to be shut down due to environmental restrictions related to air quality [37].

Estonia is integrated into the Nord Pool spot trading market. Unless government strategies change, in 2025 Estonia would become dependent on international cooperation to supply a stable source to complement its intermittent wind resources, thus reducing its energy security. In addition, according to the World Energy Council, since wind resources in Estonia and its neighbors are highly correlated, geographic integration will not fully mitigate potential intermittency problems, making the case for new energy storage capacity [36]. This future projection could be a catalyst to spur energy storage R&D for domestic purposes in Estonia.

In Singapore, on the other hand, the government has a stated focus on energy storage, with Singapore's Energy Market Authority having started an Energy Storage Program in 2015 to improve the stability of Singapore's power system and included the launch of its Energy Storage System (ESS) testbed in October 2017 [38]. Singapore currently relies on natural gas for 95% of its electricity needs, but has a national target to deploy 350 MWp of solar PV by 2020, which is the most promising renewable source available in the country [39]. There are no hydro resources, wind speeds and mean tidal range are low, and geothermal energy is not economically viable [40].

At the end of Q1 2018, Singapore had 115 MW of installed solar capacity making up approximately 0.8% of Singapore's total energy mix [40]. This is a tangible step towards the national target for 2020, reflecting Singapore's commitment to solar PV.

According to the Economic Research Institute for ASEAN and East Asia, Singapore's solar PV mix will increase to 8% of Singapore's energy mix by 2030 [39]. A study by the Sustainable Energy Association of Singapore presents a more aggressive mix, stating that solar energy could possibly meet as much as a quarter of Singapore's energy needs in 2025 [41]. However, due to the intermittent nature of solar, there are limitations to deploying solar on a large scale to generate electricity reliably in Singapore. The ESS testbed is thus seen as an important factor in enabling solar adoption in Singapore [42]. Singapore aims to develop a niche in energy storage and batteries to first ensure domestic energy security and to further explore applicability to regional and global markets. International cooperation agreements have also been a key part for Singapore's energy storage development. Singapore's Agency for Science, Technology and Research (A*STAR) recently signed an agreement with Canada's Hydro-Quebec to establish a joint laboratory to research emerging battery technologies for electric vehicles and energy storage. In addition, the German company VDE set up a Global Energy Storage Competence Cluster (GECC) in partnership with Singapore's Nanyang Technological University.

4.2.2 Regional markets for energy storage

Both countries, as export-oriented economies, are often considered gateways to Europe (Estonia) and Asia (SEA and China) (Singapore) [43]. Both regions have strong potential as markets for energy storage solutions in the future. Despite a lack of immediate focus on energy storage domestically in Estonia, the country is motivated to fund R&D and develop capacity in energy technology priorities for the European Union (EU), one of which is energy storage [34]. In fact, many Estonian energy startups take advantage of EU R&D money through programs such as Horizon 2020 funding. The EU has a stated focus on energy storage solutions and a growing need for flexibility in the energy system, which would benefit from innovation in emerging storage solutions. The EU promotes battery storage technologies through its Horizon 2020 program and also has a specific program focused on fuel cells and hydrogen development [44]. According to the European Market Monitor on Energy Storage, Europe's energy storage market grew by 49% in 2017 [45].

While grid operators in Europe still have not fully defined the best way to integrate energy storage into their business models, the UK and Germany are currently the largest markets for energy storage in Europe, with favorable policies and regulations that provide flexibility to adjust to a quickly changing market. As more customers adopt storage technologies and costs start to go down, behind-the-meter energy storage is also seeing growth, with commercial and industrial (C&I) expected to grow 45% in 2018, and was recently dubbed the most exciting segment of the European energy storage market [45]. In the C&I sector, for example, there is a lot more flexibility to use energy storage solutions that provide resilience and independence in a company's energy generation, which impacts a business' bottom line and de-risks its energy costs by not fully relying on merchant revenues of short-term grid services contracts [45]. Innovative business models, such as storage-as-a-service, are also emerging, which will make access to energy storage easier and simpler for customers and grow customer demand for energy storage-as-a-service providers.

In SEA, frequent brown outs provide a strong incentive for the reliability that renewables-plus-storage, specifically solar-plus-storage, could provide [46]. While the regulatory environment in SEA is not currently friendly to renewables-plus-storage solutions, Singapore provides the perfect opportunity to act as both a technology and regulatory testbed for future deployment in SEA countries. Currently in the Asia-Pacific region, there is 1784 MW of energy storage system capacity in the pipeline, primarily from pumped hydro [47]. The remaining battery solutions are primarily lithium ion storage projects based in the Philippines and China. The lack of a more cohesive set of regional market mechanisms and policies in the SEA region indicates there is an opportunity for Singapore to take the lead in the energy storage sector. In addition, due to regulatory barriers in other countries, Singapore can also use its regulatory sandbox as a model for other SEA countries who are interested in R&D, demonstration and deployment of energy storage technologies.

Projected energy storage deployments by market show significant growth over the next 6 years in the East Asia & Pacific, signaling room for additional R&D and innovation in energy storage technologies. Increased focus on storage is evident in the growth of initiatives such as the ASEAN Solar + Energy Storage Congress and

Expo, which will bring together developers and investors to establish a sustainable and viable business model in SEA to prepare for the future energy transition [48].

4.2.3 Why Estonia and Singapore?

Significant focus on intermittent renewable energies (wind in Estonia, solar in Singapore) in both countries' futures provides a catalyst for energy storage R&D both domestically and regionally. The two countries have separate, but similar concerns regarding current and future energy security and dependence on neighboring countries as their shares of renewable energy grows. Domestic and regional market mechanisms, such as climate and renewable targets and policies, drive a need for innovation in energy storage and energy technology in general. While Estonia is looking to reduce its energy intensity, it's not concerned in the near-term with deploying energy storage domestically for energy security purposes. By participating in joint technology innovation with Singapore, Estonia can start preparing for a future that will most likely require some form of energy storage to ensure energy independence.

Together, Singapore and Estonia can provide each other with significant development opportunities and access to regional markets by leveraging each other's strengths and resources to jointly conduct energy storage R&D. Both countries rank fairly highly in entrepreneurship, innovation, and development indicators, such as ease of doing business, enabling trade, and the global competition, innovation and entrepreneurship indices. For example, Estonia's e-residency program and Enterprise Estonia office in Singapore, along with other key stakeholders can facilitate collaboration, while convenient access to regional markets can attract potential funding sources for joint collaboration. In addition, Estonia's recent presidency of the Council of the EU in 2017 provides it with the political know-how and relationships that Estonia can employ to help Singapore navigate Europe's policy and regulations [49].

In addition, as small countries rooted in knowledge economies, both countries have energy technology R&D systems and regulatory structures that can foster innovation and growth in the energy storage innovation pipeline. Estonia's energy storage R&D capabilities tend to be in the technical capabilities present in universities, startups, and the private sector, primarily focused on export, whereas Singapore's energy storage R&D is focused on the application of technologies in its government sponsored testbed, along with select international cooperation agreements between governments and university organizations. The alignment of technical focus and scientific capabilities of Estonia and Singapore in combination with the ability to rapidly implement potential storage solutions in the Singapore testbed complement each other in developing energy storage solutions. Such solutions not only serve to support both wind and solar technologies but may, as well, find application with other diverse energy generation technologies being explored in the testbed. Successful technology demonstrations resulting from the Estonian-Singapore collaboration may find ready funding for commercial scale-up through Estonia's access to EU funding, domestic venture capital funding, and startup culture and result in the desired creation of energy storage capacity and expertise.

4.3 Proposal for joint energy storage innovation collaboration

This section will outline the proposed joint collaboration between Estonia and Singapore on energy storage innovation by first laying out a proposed plan including key 'areas of interest' within energy storage and define key stakeholders.

Second, it will identify market mechanisms across both countries and propose an IP sharing cross-license model for the collaboration.

4.3.1 Proposed joint collaboration model

Given the interest in both countries for innovation in energy, the incentive to peruse energy storage, and opportunities for a mutually beneficial partnership as outlined in the above section, we propose Estonia and Singapore establish an initial joint collaboration, jointly led by Tehnopol Science & Technology Park in Estonia and Singapore's Agency for Science, Technology and Research (A*STAR). These two entities are involved in all stages of the innovation pipeline, which will help ensure smooth handoffs and mitigate the risks of 'valley of death' between the innovation stages. Both entities receive government funding, and should work with their respective government energy R&D departments to allocate existing funding specifically to begin this collaboration. Other stakeholders, such as Enterprise Estonia, Enterprise Singapore, the National Research Council of Estonia, and the Energy Market Authority in Singapore, can play key supporting roles.

The two organizations will collaborate to identify a joint development program with a realistic roadmap and solicitation process to both receive promising joint R&D collaboration research proposals and define opportunities to demonstrate, deploy and commercialize existing energy storage innovations, all of which will tie into the joint innovation pipeline. While initial funding will come from the two entities and ultimately government funds, Tehnopol and A*STAR will also help procure funding from any of the key stakeholders, such as the EU's Horizon 2020 fund, Estonian venture capitalists, or Singaporean government entities. The two entities will also be supported by resources from both sides, such as supporting mentorship and commercialization and market opportunities that are identified from each countries' respective enterprise and business development offices. For example, promising products or startups that come out of the joint collaboration could be candidates for inclusion in Tehnopol's startup incubator or use Singapore's ESS testbed to test product capabilities.

The joint collaboration will include four main components to accelerate successful development:

1. Creation of the Singapore-Estonia Energy Storage Collaboration Initiative (SEES-CI) (see **Figure 1**), which will focus on bringing together Singaporean and Estonian startups, university research institutes, and government R&D programs focused on energy storage technologies in both Estonia and Singapore.

    - The Initiative will be housed within a new Singapore-Estonia office for Energy Storage Collaboration in Tehnopol, the Baltic region's largest science and technology park, based in Tallinn. This office can eventually house an Enterprise Singapore office and serve as the country's Baltic and North/Eastern European headquarters, where it currently has no presence [50].

    - The SEES-CI will plan up front for success, so that IP cross-license agreements and commercial out-licenses and in-licenses necessary to commercial success of the initiative will be anticipated up-front and managed collaboratively by Enterprise Estonia and Enterprise Singapore (see section on Intellectual Property Sharing Model for Joint Energy Storage Technology Innovation).

2. Host an Annual Singapore-Estonia Energy Storage Expo & Workshop (alternately hosted each year) to interact and establish more defined areas for energy

research and innovation, as well as identify opportunities for demonstration in testbeds in either country, and identify commercialization/market opportunities and supply chain development throughout Europe and/or Southeast Asia.

- One outcome of the annual workshop would be to identify business delegations that could visit the other country, sponsored by Enterprise Estonia and Enterprise Singapore. The delegation visits would include visiting key stakeholders, including universities, R&D centers, startups, demonstration sites, and energy storage related government departments.

- The annual workshop could also serve as the kick-off event for the next annual round of requests for proposals solicited by the SEES-CI, in particular the call for joint R&D proposals noted as key element 4 below.

3. Establish a Resource Sharing Agreement in which Estonia and Singapore agree to share non-financial resources that enable energy technology development and clearly define these resources and who and how the resources will be applied consistent with the CEES-CI program. 'Resources' could include limited access to Singapore's ESS for demonstration, mentorship and SME time, laboratory researchers and materials, etc.

4. Call for joint R&D proposals between Estonia and Singapore in key 'areas of interest' related to energy storage. The proposals would require plans for demonstration and deployment, as well as a market analysis. Funding for these proposals could be set up with initial funding from both governments, with the potential for additional grants or funding from other sources such as the EU Horizon 2020 fund, Tehnopol, or other key stakeholders. The first three proposed 'areas of interest' include:

   - Hybrid Energy Storage Solutions (HESS)

   - Hydrogen Energy Storage and Fuel Cells

   - Supercapacitors

4.3.2 Key stakeholders and collaborators

The key energy innovation stakeholders in a joint energy storage innovation collaboration between Singapore and Estonia will include government organizations, universities, companies and startups, industry organizations, and regional institutions.

4.3.2.1 Government organizations

The primary government organizations involved in joint energy technology innovation and energy storage technology is the Ministry of Trade and Industry and Ministry of Foreign Affairs in Singapore and the Ministry of Economic Affairs and Communication and Ministry of Education and Science in Estonia. In this joint collaboration scheme, we propose that A*STAR, which falls under Singapore's Ministry of Trade and Industry and considers itself a 'catalyst, enabler and convener of significant research initiatives among the research community in Singapore and beyond' to co-lead the innovation collaboration with Tehnopol Science & Technology Park in Tallinn, Estonia [51].

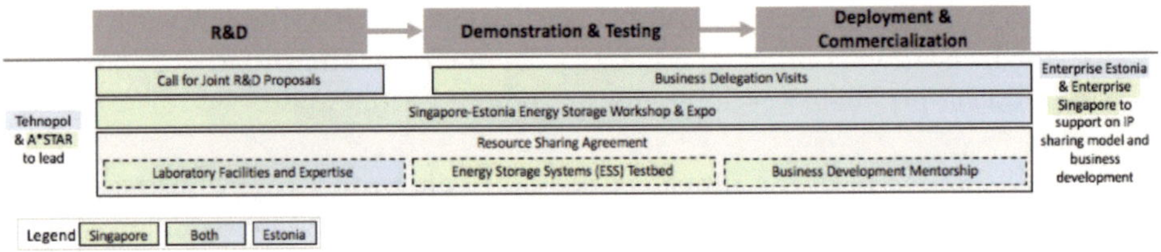

Figure 1.
Proposed Singapore-Estonia energy storage collaboration initiative (created by author).

Tehnopol is a research and business science park that includes a startup incubator and is a public-private-partnership (PPP) that receives funding from the city of Tallinn, Tallinn Technical University (TTU), and the central government of Estonia [34, 52]. Tehnopol has three main stages of support: (1) Prototron Fund, which is seed funding for an idea and without requiring equity from any future startup or business; (2) startup incubator with mentorship network and guidance; and, (3) business development and export assistance for companies with mature products [34]. Tehnopol often works to identify university research teams developing valuable technology, and begins mentoring them through the Prototron Fund, which can lead to establishing a startup.

These two institutions, A*STAR and Tehnopol are both focused on developing the full innovation pipeline with public and private partners, which will allow them to be creative in identifying joint collaboration opportunities and leverage partnerships across the full spectrum of stakeholders. Thus, they will be able to make connections between more R&D focused organizations such as universities, the ESS test bed and Estonian research council and organizations focused more on business development, such as Enterprise Estonia and Enterprise Singapore.

Each of these entities have vested interests in identifying business, innovation, and talent opportunities that drive R&D, innovation and commercialization within their respective countries–and will be able to do the same through joint collaboration.

4.3.2.2 Universities

Estonia and Singapore have strong tertiary education, with the Innovation Index ranking Singapore 1st and Estonia 27th out of 126 countries [53]. Each country has a few main universities that focus on science, technology and energy research, and are eligible for funding for research grants within the state and regionally. In Estonia, each of the universities identified in the stakeholder map have energy related programs. In addition, university R&D has a pathway to further develop its prototypes, through Tehnopol or its strong culture of startups and venture capital. With the shared TTU and Tehnopol campus, researchers at TTU have an opportunity to work with startups and entrepreneurs to further explore their idea. In addition, Tehnopol often serves as a 'testbed' for its members' technologies, such as a pizza-delivering robot and solar energy street lights [34]. They are also eligible to apply for research grants from the Estonian Research Council and other regional EU funds, such as Horizon 2020 funding.

In Singapore, each major university has a program or research focused on energy and on energy storage in particular, and are eligible to apply for funding for A*STAR or EMA's call for research proposals. In fact, the National University of Singapore and Nanyang Technological University and will be launching a Singapore Energy Center in partnership with ExxonMobil in early 2019 to explore innovative ideas and develop talent to meet future energy needs in Asia Pacific [54].

4.3.2.3 Companies and startups

In energy storage technology, a few key startups in each country could be key elements in a Singapore-Estonia collaboration. Through joint collaboration, they could further R&D on their projects, test innovations in the ESS testbed or other demonstration sites, or deploy their products to new markets. For example, SkeletonTech, an Estonian supercapacitor startup, originally received funding from an Estonian-Norwegian joint energy technology innovation program in 2013 and has received at least one round of venture capital funding [25]. However, it is focused primarily on the European market, and could use the Singapore-Estonia joint innovation program as an opportunity to explore new markets and test its technologies in Singapore's climate.

4.3.2.4 Industry organizations

The industry organizations in Estonia and Singapore are focused mostly on renewable energy, with one industry organization in Estonia focused on hydrogen energy. These associations can play a role in helping advocate for regulatory and policy mechanisms that will promote renewable energy and ultimately energy storage and also work with organizations such as Enterprise Singapore and Enterprise Estonia in developing a strong community of partners within each country. As renewable energy becomes more prevalent in both countries, industry organizations will become more and more important for building up skills capacity in energy storage.

4.3.2.5 Regional institutions

As primarily export markets, both Singapore and Estonia have interests in exporting energy technologies to the Southeast Asian and European markets. Both regions host large energy storage conferences, which bring together startups, researchers, government officials, and experts, and could be a good way to advance opportunities between an Estonia and Singapore cooperation. In Asia, the Asia Energy Storage Association was recently founded in August 2018 and aims to be a regional platform for all energy storage industry stakeholders to promote the best interests of the energy storage sector in Asia [55].

The ASEAN Energy Business Forum brings together Ministers of Energy from ASEAN countries to identify business opportunities in the region. However, currently most energy storage research and demonstration in the region is on a country by country basis, with little clear research funding from a regional perspective (ASEAN or otherwise) [56]. In Europe, funds such as Horizon2020 and PowerUp! have been a crucial part in advancing Europe's energy priorities, including energy storage. Europe also separates hydrogen and fuel cell research from its other energy storage goals, and has separate funding in both areas as well.

4.4 Market mechanisms for energy storage in Singapore and Estonia

This section will discuss how market mechanisms for energy storage will be divided between Singapore, Estonia and the EU. While the SEA region has very little regional funding or policies to support energy storage, Singapore can help lead the efforts in identifying key country markets in SEA in the absence of a more cohesive regional market mechanism.

Because both countries are relatively small in population and area, neither country will be able to sustain a market on its own for any given technology.

However, Estonia and Singapore, respectively, have still maintained levels of R&D capacity, skills development and capacity building, and innovation by leveraging their positions as "gateways" to Europe and Southeast Asia. Both countries also have strong abilities to export high-tech products and services [57]. Singapore especially has become a key location for joint collaboration in energy technology, and more recently energy storage technology [38].

The below chart (**Figure 2**) outlines the major market push and pull mechanisms in Singapore, Estonia, and the EU. From a SEA perspective, while the market for energy storage has the potential for extreme growth, the region does not have any cohesive mechanisms (policies, funding, etc.) that provide pull or push for energy storage. However, individual countries such as the Philippines, do have market demand for energy storage technologies, and new forums such as the ASEAN Solar + Energy Storage Congress and Expo will start to identify deployment opportunities in the SEA market. Together, these policies, funds, and market demand, combined with the innovation pipeline and stakeholder structure in Estonia, Singapore, and the EU provide a model for success in Estonian-Singaporean energy storage collaboration.

4.4.1 Market push

From a funding & services perspective, both countries as well as the EU offer funding opportunities to conduct R&D activities related to energy storage through programs such as the Estonian Research Council Grants, Horizon 2020 funding, and A*STAR grants and funding. For example, the EU's program on Accelerating Clean Energy Innovation plans to deploy €2 billion towards its priority R&D areas, one of which is 'developing affordable and integrated energy storage.' [58] The EU Emission Trading Scheme (ETS) Innovation Fund was created in 2015 and creates a fund from ETS revenues to support large-scale demonstration of energy innovation, including energy storage [59]. In addition, Singapore's Energy Storage Program, which includes the ESS testbed initiative, and Estonia's Tehnopol Science & Technology park, offer programs such as the Prototron Fund and Startup Incubator that will push R&D in energy storage sector.

One interesting item to note is that both Singapore and Estonia had formal energy technology innovation programs and funding, but did not renew them

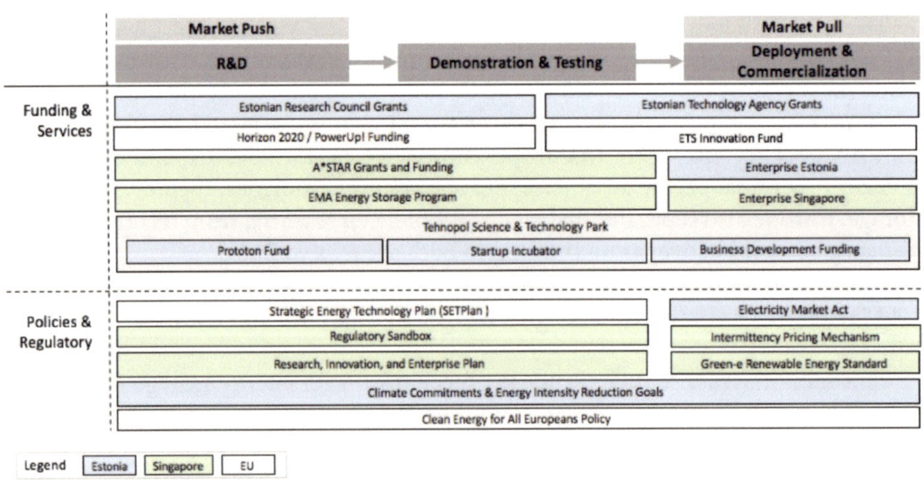

Figure 2.
Market push and pull for energy storage technology in Estonia, Singapore, and the EU (Created by author).

after their five-year periods. Singapore had an Energy Innovation Program Office (EIPO), which is no longer in operation, and allocated S$195 million from the National Research Foundation to promote R&D in the energy sector from 2010 to 2015 [60]. The Estonian Energy Technology Program also had €6.5–9.5 million allocated from 2008 to 2013, which supported R&D, technology transfer, and joint activities and value added services [44]. With neither of these mechanisms in place anymore, both countries will have to leverage alternative sources of funding.

In the policy and regulatory space, Singapore provides more mechanisms to encourage R&D and deployment & commercialization of energy storage technology. For example, Singapore's regulatory sandbox provides impetus to innovate and test out new technologies without the risk of regulatory barriers. In addition, its Research, Innovation and Enterprise Plan has set aside $275 million USD for key research, development and deployment initiatives in the energy space, focusing on areas such as solar, energy storage, smart grids, and green buildings [40]. Singapore is also planning to implement a carbon tax by 2025, which could provide revenue that will induce additional innovation in low carbon energy technologies and energy storage [61]. In Estonia, its Estonian National Development Plan of the Energy Sector for 2030 (ENMAK 2030+) and climate policy principles for 2050 (KPP) incorporate climate commitments as part of the EU's Paris Accord commitments, energy intensity reduction goals, and a shift away from shale power generation to wind and renewables [37]. These factors combined with a future concern for energy security and lack of large-scale storage solutions to address intermittency of renewables will help push energy storage R&D. Finally, the EU's Strategic Energy Technology Plan, which focuses on accelerating development of energy technologies, along with the Clean Energy for All Europeans package, which proposes EU policies and standards on several aspects of energy including renewables, efficiency, and electricity markets, will both provide regulatory certainty and encourage essential investments into the energy storage sector [62, 63].

4.4.2 Market pull

From a funding & services perspective, the Estonian Technology Agency provides grants that assist in the piloting and deployment of new technologies. Tehnopol also provides business development funding and assistance. Finally, Enterprise Estonia and Enterprise Singapore, which are business development organizations in each country, promote business and regional policy and are critical parts of their respective national support systems for entrepreneurship and product commercialization. They also provide training opportunities and connect public sector, research institutions and private enterprise to help commercialize products in export markets [64].

In the policy and regulatory space, the EU's Clean Energy for All Europeans package provides binding standards on renewables and greenhouse gas emission reductions that will help create market pull (and push) for energy storage technologies. Estonia's climate policy as mentioned in the market push section, such as its commitment to reduce energy intensity and shut down shale energy generation plants, could also create a market pull for energy storage. Estonia also recently passed the Electricity Market Act, which introduces technology-neutral auctions for renewable energy projects, includes a clause that states that the Competition Authority can impose obligations to invite tenders for energy storage devices [65]. Singapore recently announced an intermittency pricing mechanism (IPM) that will require reserves to be available to ensure stability of supply and could be a significant market pull to deploy energy storage solutions in Singapore's grid [29]. Finally, Singapore's Green-e Renewable Energy Standard (RES) certifies renewable energy products and aims to accelerate renewable electricity markets and develop a mechanism through which consumers can demand renewable electricity. This policy could also increase demand for energy storage to mitigate intermittency [66].

4.5 Intellectual property sharing model for joint energy storage technology innovation

With regard to intellectual property (IP), Singapore has a somewhat stronger IP and innovation rating while Estonia is somewhat weaker in this area. It is especially important for the joint proposal to outline how IP created in the technical collaboration will be shared to mitigate risks associated with jointly sharing IP rights and provide the fairest model [67]. Both Singapore and Estonia have know-how related to their respective regional markets and policies that can help leverage the collaboration on IP and guide the introduction of successful technologies into nearby SEA and European markets, respectively.

It is proposed that, consistent with commercial collaborations, Singapore and Estonia agree at the outset to a royalty-free cross-license of the intellectual property developed in collaboration–free for each country to deploy in their respective countries plus a sharing of returns on such intellectual property if licensed outside their countries. This mutual commitment would encourage the two countries to provide the complementary technical, financial and management resources needed for success. Such intellectual property collaboration is enhanced by the geographic separation of Singapore and Estonia which limits concerns about competition between their economies.

To encourage company, start-up and university participation, the SEES-CI would require that before third-party technology is used in the collaboration by a country, the third-party licenses or makes available its technology for such collaborative research activities and commits, if requested, to non-exclusively license its IP on commercially reasonable terms and conditions through the organization chosen by Estonia and Singapore to commercialize the collaboration IP outside their borders [68].

Cross-licensing between Estonia and Singapore and the promise of commercial licenses are a priori commitments that will short-cut the win-lose thinking that otherwise dominates many collaborations, where the focus is misplaced from desired outcomes and drifts to questions about sole or joint inventorship, and the withholding of ideas arising from concern over legal rights. Rather, it is intended to foster true collaboration and the free flow of ideas.

For intellectual property to qualify as falling under the joint cooperation program, the SEES-CI must have agreed to the project proposal and provided some of the funding as part of SEES-CI, and have agreed to the scope of work and the deliverables of the project. This will make clear the scope of technology to be captured under the agreement and that will be cross-licensed to Estonia and Singapore, and available through the organization chosen by Estonia and Singapore to commercialize the collaboration IP outside their borders.

Key individuals from each country from the designated institutions may also be identified and would be expected to exhibit the leadership and integrity to accurately manage the disclosure of contributions to the development and innovation process in performance of projects under the collaboration, which could include: conception of ideas; material contributions to the development of an invention, data, information, software, hardware or trade secrets; providing solutions or troubleshooting to problems; implementing the invention; and/or providing ordinary assistance, such as performing routine tasks or executing testing. All depends on SEES-CI managing defined tasks and responsibilities assigned to each country in a fully thought-out roadmap of funding and technology development, implemented through a series of projects performed by government, university and private stakeholders.

Singapore and Estonia will be responsible for patenting the technology in their respective countries and can in the course of defining the collaboration, determine details of cross-licensing as previously discussed, or decide to proceed as co-owners of the patents [69]. All costs related to filing, renewal and prosecution fees related to the IP will be shared equally. Where private research efforts, universities, or non-government laboratories are part of the collaboration on behalf of Singapore or Estonia, the countries may wish to adopt an approach much like the Bayh-Dole rights in the United States, under which the funding government receives non-exclusive rights which may be shared with the other collaborating country, and the non-governmental collaborating entity may elect to pursue IP protection for commercialization [70].

As between Estonia and Singapore, a commercial-like arrangement can be adopted should one country wish to file for patent protection in a country where the other does not (i.e. if Singapore wants to file in China and Estonia does not), the other party will have the right to file and maintain the patent or patent application at its own expense. In that case, the non-participating country would not share in any income from the country where they do not support or withdraw support of the patent costs [43].

It is proposed that Estonia and Singapore agree to license collaboration intellectual property outside their countries through a single entity acting on their behalf who could also bundle any enabling privately held IP rights into a licensing package. Regardless, as is typical, no obligation would exist to share licensing income for pre-existing background intellectual property or inventions that are developed outside the collaboration, except as may be agreed.

Patent enforcement strategies will also be agreed upon as part of the collaboration. Should a dispute arise between Singapore and Estonia that cannot be resolved by mutual agreement of parties, the parties would decide to work with the World Intellectual Property Organization or another international body they may designate to determine the correct course of action for arbitration and mediation [71].

As business development institutions, Enterprise Estonia and Enterprise Singapore can work with Tehnopol and A*STAR to help monitor the IP sharing agreement and help identify appropriate markets for deployment and commercialization. By allowing the Enterprise arms of Singapore and Estonia to lead the IP sharing agreement, both entities will be motivated to not only ensure proper protection of innovation, but also ensure fair and transparent IP sharing agreements within the two countries and internationally.

5. Conclusions

With environmental challenges increasingly becoming global challenges that transcend boundaries, so too will opportunities for joint collaboration in energy innovation. In the words of Louis Pasteur, "Science knows no country, because knowledge belongs to humanity, and is the torch which illuminates the world [72]." As international collaboration continues to grow within the context of a more globalized system of science and innovation, the role for small country joint collaboration will grow stronger.

Small country joint innovation can help address some of the unique challenges that small countries face and develop balanced and mutually beneficial collaboration. This paper discusses challenges, criteria, benefits, and enablers for successful small country joint innovation. Together, small countries can leverage each other's strengths to build successful and creative models of innovation collaboration. These models can be valuable for countries big and small, developed and developing, and provide one more tool for combating energy issues around the world. A proposed joint collaboration between Singapore and Estonia presents a model for energy storage technology, that, if successful, could continue to evolve, leading to increasing opportunity for win-win collaboration across a spectrum of energy opportunities.

While the case study in this paper uses energy storage technology as a basis for small country joint innovation, storage is just one of a suite of potential innovations that a small country could pursue to make choices around energy investments. Depending on a given country's energy needs, access to investment, and innovation capabilities, energy technologies such as solar PV, energy storage, fuel cells, geothermal, wind, biomass and biofuels are all additional areas in which small country could pursue joint innovation. Innovation in these areas can drive decisions around investments in additional areas such as grid management software and energy infrastructure, which can eventually lead to development at scale and a replicable model for other countries.

In addition, the opportunity for small and large country innovation can be further enhanced by 'polycentric innovation.' The idea of polycentric innovation, which consists of shared networks of international talent, capital, and ideas that initially take advantage of regional opportunities and are then integrated into global innovation networks, can be an important concept for small country joint innovation [73]. To enhance policy effectiveness in institutional development and optimize processes, Nyangon & Byrne (2018) and Liu & Liu (2018) argue that nations should mainstream polycentric innovation at multiscale levels to advance greater technological innovation ecosystem, customer enterprises, and business model innovation [74, 75].

Acknowledgements

The authors would like to thank Professor Deborah Bleviss, former Administrative Director of the Energy, Resources and Environment Program at the Johns Hopkins School of Advanced International Studies, for her guidance and support in the writing of this chapter.

Author details

Anneliese Gegenheimer[1*] and Charles Michael Gegenheimer[2†]

1 Johns Hopkins School of Advanced International Studies, Washington, DC, USA

2 Battelle Memorial Institute, Columbus, OH, USA

*Address all correspondence to: anneliese.gegenheimer@gmail.com

†Retired.

IntechOpen

© 2020 The Author(s). Licensee IntechOpen. Distributed under the terms of the Creative Commons Attribution - NonCommercial 4.0 License (https://creativecommons.org/licenses/by-nc/4.0/), which permits use, distribution and reproduction for non-commercial purposes, provided the original is properly cited. (cc) BY-NC

References

[1] Collaborate to innovate [Editorial]. Nature Energy. 2017;**2**(11):831. DOI: 10.1038/s41560-017-0046-3

[2] World Health Organization. Climate Change Increasingly Affects Small Countries [Internet]. 2018. Available from: http://www.euro.who.int/en/countries/iceland/news/news/2018/6/climate-change-increasingly-affects-small-countries [Accessed: July 29, 2019]

[3] Morford S. Climate Change Will Cost U.S. More in Economic Damage Than Any Other Country But One [Internet]. 2018. Available from: https://insideclimatenews.org/news/24092018/climate-change-economic-damage-america-social-cost-carbon-china-india-russia [Accessed: July 29, 2019]

[4] Council of Foreign Relations. Harnessing international cooperation to advance energy innovation [Internet]. 2017. Available from: https://www.cfr.org/report/harnessing-international-cooperation-advance-energy-innovation [Accessed: July 29, 2019]

[5] Kamalski J. Small countries lead international collaboration [Internet]. Research Trends. 2009;**14**. Available from: https://www.researchtrends.com/issue14-december-2009/country/ [Accessed: July 29, 2019]

[6] Kundera, M. Die Weltliteratur [Internet]. The New Yorker. 2006. Available from: https://www.newyorker.com/magazine/2007/01/08/die-weltliteratur [Accessed: July 29, 2019]

[7] Lanvin B. The World's Most Innovative Countries [Internet], 2018. Available from: https://knowledge.insead.edu/entrepreneurship/the-worlds-most-innovative-countries-2018-9666#c24bMHBqPjxuP35Q.99 [Accessed: July 29, 2019]

[8] Radu S. The Fad Behind the Tag [Internet]. U.S. News & World Report. 2019. Available from: https://www.usnews.com/news/best-countries/articles/2019-01-23/why-countries-care-about-being-innovative [Accessed: July 29, 2019]

[9] Rossetti P. No Country Can Fix Climate Change on Its Own [Internet]. American Action Forum. 2019. Available from: https://www.americanactionforum.org/insight/primer-us-cant-fix-climate-change-on-its-own/ [Accessed: July 29, 2019]

[10] United Nations Development Programme. Even a Small Country Can Bring a Big Change in the World [Internet]. 2018. Available from: https://medium.com/@UNDP/even-a-small-country-can-bring-a-big-change-in-the-world-cb7c6654fe6b [Accessed: July 29, 2019]

[11] Kreisberg D, Carrasco N, Jordy D, and A. Giardino. Learning from Big Innovations in Small Island States [Internet]. World Bank. 2018. Available from: https://medium.com/world-of-opportunity/learning-from-big-innovations-in-small-island-states-371f50d3489e [Accessed: July 29, 2019]

[12] Tawney L. Developing Countries Can Spark Wave of Innovation in Clean Energy [Internet]. World Resources Institute. 2011. Available from: https://www.wri.org/blog/2011/06/developing-countries-can-spark-wave-innovation-clean-energy [Accessed: July 29, 2019]

[13] Chawla L. International Collaborations Growing Fast [Internet]. Nature Index. 2018. Available from: https://www.natureindex.com/news-blog/international-collaborations-growing-exponentially [Accessed: July 29, 2019]

[14] Bergseng A. Climate Change Action in Small Countries and

Regions [Internet]. ClimateXchange. 2019. Available from: https://www.climatexchange.org.uk/blog/climate-change-action-in-small-countries-and-regions/ [Accessed: July 29, 2019]

[15] Kattel R, Randma-Liiv T, Kalvet T. Small states, innovation and administrative capacity. In: Bekkers V, Edelenbos J, Steijn B, editors. Innovation in the Public Sector. London: Palgrave Macmillan; 2010. pp. 61-81. DOI: 10.1057/9780230307520_4

[16] Arunachalam S, Doss M. Science in a small country at a time of globalization: Domestic and international collaboration in new biology research in Israel. Journal of Information Science. 2000;**26**(1):39-49. DOI: 10.1177/016555150002600104

[17] Rae J, Westlake S. When Small is Beautiful: Lessons Learned from Highly-Innovative Smaller Countries [Internet]. Nesta. 2014. https://media.nesta.org.uk/documents/when20small20is20beautiful20final.pdf [Accessed: July 29, 2019]

[18] Brown A. The Pitfalls of Going from Pilot to Scale, or Why Ecological Validity Matters [Internet]. International Initiative for Impact Evaluation. 2016. Available from: https://www.3ieimpact.org/blogs/pitfalls-going-pilot-scale-or-why-ecological-validity-matters [Accessed: July 29, 2019]

[19] Zanotto S, Haeffner C, Guimaraes J. Unbalanced international collaboration affects adversely the usefulness of countries scientific output as well as their technological and social impact. Scientometrics. 2016;**109**(3):1789-1814. DOI: 10.1007/s11192-016-2126-8

[20] Edquist, C, Hommen, L. editros. Small Country Innovation Systems. Cheltenham: Edward Elgar Publishing Limited, 2008

[21] Bai X, Liu Y. International collaboration patterns and effecting factors of emerging technologies. PLoS One. 2016;**11**(12):e0167772. DOI: 10.1371/journal.pone.0167772

[22] United Nations Educational, Scientific and Cultural Organization. International Scientific Collaboration has Become a Must, Says Report [Internet]. 2017. Available from: http://www.unesco.org/new/en/media-services/single-view/news/international_scientific_collaboration_has_become_a_must_sa/ [Accessed: July 29, 2019]

[23] World Competitiveness Center. European Economies Best at Recruiting and Retaining Top Talent, Says IMD Study. International Institute of Management Development [Internet]. 2017. Available from: https://www.imd.org/news/updates/european-economies-best-at-recruiting-and-retaining-top-talent-says-imd-study/ [Accessed: July 29, 2019]

[24] Baker S. Ignoring Developing Countries. Inside Higher Ed [Internet]. 2018. Available from: https://www.insidehighered.com/news/2018/09/21/study-finds-limited-collaboration-between-research-elites-and-developing-nations [Accessed: July 29, 2019]

[25] Skeleton Technologies. Blog, News and Events. Available from: https://www.skeletontech.com/news/page/13 [Accessed: July 29, 2019]

[26] Chakravorti B, Bhalla A, Chaturvedi R. Which Countries are Leading the Data Economy? Harvard Business Review [Internet]. 2019. Available from: https://hbr.org/2019/01/which-countries-are-leading-the-data-economy [Accessed: July 29, 2019]

[27] United Nations. Collaboration, Solidarity Mindset among Developing Countries Thriving Amid Decline in Aid from North. 2019. Available from: https://www.un.org/press/en/2019/

dev3390.doc.htm [Accessed: July 29, 2019]

[28] Pilar K. Personal Email with Pilar K, Project Manager, Energy Department, Estonian Ministry of Economic Affairs and Communications; 8 November 2019

[29] Energy Market Authority. Energy Storage System [Internet]. Singapore Government. Available from: https://www.ema.gov.sg/Energy_Storage%20System.aspx [Accessed: December 01, 2018]

[30] Enterprise Singapore. Singapore Industry Capabilities [Internet]. Available from: https://ie.enterprisesg.gov.sg/Partner-Singapore/Singapore-Industry-Capabilities/ [Accessed: December 01, 2018]

[31] Invest in Estonia. Estonia is Increasing Its Visibility in Southeast Asia [Internet]. 2016. Available from: https://investinestonia.com/estonia-is-increasing-its-visibility-in-southeast-asia/

[32] Tegos M. Estonia's E-Residency Program Makes It Easy for Singaporeans to do Business in the EU. Tech in Asia [Internet]. 2016. Available from: https://www.techinasia.com/estonia-e-residency-singapore-entrepreneurship [Accessed: December 01, 2018]

[33] Silver Tambur. Estonia Delivers a Cybersecurity Training to Singapore. Estonian World [Internet]. 2018. Available from: http://estonianworld.com/security/estonia-delivers-a-cybersecurity-training-to-singapore/ [Accessed: December 01, 2018]

[34] Saksing R. Personal Interview with Ragmar Saksing, Greentech Sector Manager, Tallinn Science Park Tehnopol; 5 December 2018

[35] U.S. Department of Commerce. Estonia Country Commercial Guide [Internet]. 2017. Available from: https://www.export.gov/article?id=Estonia-Energy [Accessed: December 01, 2018]

[36] Bulakh A, Tuohy E, Kearns J. Impacts of Climate Policy on Estonian Energy Security. Tallinn: World Energy Council; 2016. 11p. Available from: http://www.wec-estonia.ee/documents/83/ICDS_climate_report_2016-03-23.pdf

[37] Lindroos T, Lehtila A, Koljonen T, Kofeoe-Wiuff A, Hethey J, Dupont N, et al. Baltic Energy Technology Scenarios 2018. Copenhagen: Nordic Council of Ministers; 2018. 161p. Available from: http://norden.diva-portal.org/smash/get/diva2:1195548/FULLTEXT01.pdf

[38] EDB Singapore. Energy Storage's Day in the Singapore Sun. Eco-Business [Internet]. 2018. Available from: https://www.eco-business.com/news/energy-storages-day-in-the-singapore-sun/ [Accessed: December 01, 2018]

[39] Loi Allen and Jacqueline Tao. Singapore Country Report. Energy Outlook and Energy Saving Potential in East Asia 2016. ERIA Research Project Report 5. 2016. Jakarta: ERIA. 298p. Available from: http://www.eria.org/RPR_FY2015_No.5_Chapter_15.pdf

[40] Energy Market Authority. Renewable Energy Overview [Internet]. Singapore Government. 2018. Available from: https://www.ema.gov.sg/renewable_energy_overview.aspx [Accessed: December 01, 2018]

[41] Low I. The Future for Singapore's Energy Burns Bright [Internet]. Channel News Asia. 2017. Available from: https://home.kpmg.com/sg/en/home/media/press-contributions/2017/11/the-future-for-singapore-energy-burns-bright.html [Accessed: December 01, 2018]

[42] Tan E. Personal email with Tan E, Manager, Corporate Communications Department, Singapore Energy Market Authority; 27 November 2018

[43] Vasarais K. Personal interview with Vasarais K, President, Imtex Membrances and Low Carbon Technology Entrepreneur; 6 December 2018

[44] European Commission. Energy Storage. Research, Technology and Innovation [Internet]. n.d. Available from: https://ec.europa.eu/energy/en/topics/technology-and-innovation/energy-storage [Accessed: December 01, 2018]

[45] Colthorpe A. Commercial and Industrial is Most Exciting Part of Europe's Energy Storage Market [Internet]. Energy Storage News. 2018. Available from: https://www.energy-storage.news/news/commercial-and-industrial-is-most-exciting-part-of-europes-energy-storage-m

[46] Kenning T. Southeast Asia: Early Investors Will Gain Huge Advantage as Solar-Plus-Storage Becomes Viable [Internet]. Energy Storage World Forum. 2017. Available from: https://energystorageforum.com/news/energy-storage/will-early-solar-plus-storage-investors-gain-upper-hand-southeast-asia

[47] Gnanasagaran A. Improving Battery Storage in Southeast Asia [Internet]. The ASEAN Post. 2018. Available from: https://theaseanpost.com/index.php/article/improving-battery-storage-southeast-asia

[48] ASEAN Solar + Energy Storage Congress & Expo 2018 [Internet]. Leader Associates. 2018. Available from: http://www.leader-associates.com/asean2018/ [Accessed: December 01, 2018]

[49] Politico. Estonia's Presidency: How It Went [Internet]. 2017. Available from: https://www.politico.eu/article/estonias-presidency-how-it-went/

[50] Enterprise Singapore. Contact Us [Internet]. n.d. Available from: https://www.enterprisesg.gov.sg/contact-info/overseas-centres [Accessed: December 01, 2018]

[51] Agency for Science, Technology and Research. Overview [Internet]. n.d. Available from: https://www.a-star.edu.sg/About-A-STAR/Overview [Accessed: December 01, 2018]

[52] Tehnopol Science Park. About Us [Internet]. n.d. Available from: https://www.tehnopol.ee/en/about/ [Accessed: December 01, 2018]

[53] Global Innovation Index, Global Innovation Index Report [Internet]. 2018, INSEAD, WIPO, Cornell. Available from: https://www.globalinnovationindex.org/home [Accessed: December 01, 2018]

[54] National University of Singapore. NUS and NTU to set up Singapore Energy Centre in 2019 [Internet]. Centre for Energy Research & Technology. 2017. Available from: https://www.eng.nus.edu.sg/cert/2017/11/04/nus-and-ntu-to-set-up-singapore-energy-centre-in-2019/

[55] Asia Clean Energy Summit. Organisers & Partners. Asia Clean Energy Summit 2018 [Internet]. 2018. Available from: https://www.asiacleanenergysummit.com/about/organisers-and-partners/ [Accessed: December 01, 2018]

[56] Wong A, Musch R, Lin Z. Energy Storage–Changing and Charging the Future in Asia. Singapore: Hogan Lovells Lee & Lee; 2018. 5p. Available from: https://www.hoganlovells.com/~/media/hogan-lovells/pdf/2018/july_2018_energy-storage-in-asia.pdf

[57] Cavegn D. Analysis: Estonia Failing to Meet Research and Development Goals [Internet]. ERR.ee. 2018. Available from: https://news.err.ee/855683/analysis-estonia-failing-to-meet-research-and-development-goals

[58] Bortolotti M. editor. EASE-EERA European Energy Storage Technology Development Roadmap 2017 Update [Internet]. n.d. Available from: https://eera-es.eu/wp-content/uploads/2016/03/EASE-EERA-Storage-Technology-Development-Roadmap-2017-HR.pdf [Accessed: December 10, 2018]

[59] Ecologic Institute. The Innovation Fund: How Can It Support Low-Carbon Industry in Europe? [Internet]. 2018. Available from: https://www.ecologic.eu/15361 [Accessed: December 10, 2018]

[60] National Research Foundation. NRF Milestones [Internet]. Prime Minister's Office Singapore. n.d. Available from: https://www.nrf.gov.sg/about-nrf/national-research-foundation-singapore/nrf-milestones [Accessed: 2018-12-10]

[61] Kennedy J. How Induced Innovation Lowers the Cost of a Carbon Tax [Internet]. Information Technology & Innovation Foundation. 2018. 2p. Available from: http://www2.itif.org/2018-carbon-tax-report.pdf

[62] European Commission. Clean energy for all Europeans [Internet]. Energy Strategy and Energy Union. Available from: https://ec.europa.eu/energy/en/topics/energy-strategy-and-energy-union/clean-energy-all-europeans [Accessed: December 01, 2018]

[63] European Commission. Roundtable Underlines the Role of Energy Storage in the Clean Energy Transition. 2018. Available from: https://ec.europa.eu/info/news/roundtable-underlines-role-energy-storage-clean-energy-transition-2018-mar-01_en

[64] Enterprise Estonia. About Us. Available from: https://www.eas.ee/eas/?lang=en [Accessed: December 01, 2018]

[65] Teataja R. Electricity Market Act [Internet]. Estonian Ministry of Economic Affairs and Communications. 2014. Available from: https://www.riigiteataja.ee/en/eli/528082014005/consolide [Accessed: December 01, 2018]

[66] Energy Storage Association. Renewable Integration Benefits [Internet]. n.d. Available from: http://energystorage.org/energy-storage/energy-storage-benefits/benefit-categories/renewable-integration-benefits [Accessed: December 01, 2018]

[67] Millien R. Fully Baking Joint IP Ownership into Collaboration Agreements. IP Watchdog. 2016. Available from: http://www.ipwatchdog.com/2016/02/16/fully-baking-joint-ip-ownership-into-collaboration-agreements/id=66149/

[68] European IPR Helpdesk. Fact Sheet–IP Joint Ownership. Brussels: European Commission; 2015. pp. 1-11. Available from: https://www.iprhelpdesk.eu/sites/default/files/newsdocuments/Fact-Sheet-IP-Joint-Ownership.pdf

[69] UpCounsel. Jointly Owned Intellectual Property: Everything to Know. n.d. Available from: https://www.upcounsel.com/jointly-owned-intellectual-property [Accessed: December 12, 2018]

[70] UpCounsel. Bayh-Dole Act: Everything You Need to Know [Internet]. n.d. Available from: https://www.upcounsel.com/bayh-dole-act [Accessed: December 12, 2018]

[71] World Intellectual Property Organization. Alternative Dispute Resolution [Internet]. n.d. Available from: https://www.wipo.int/amc/en/ [Accessed: December 12, 2018]

[72] Haynes T. Science Diplomacy: Collaboration in a Rapidly Changing World [Internet]. Harvard University. 2018. Available from: http://sitn.hms.

harvard.edu/flash/2018/science-diplomacy-collaboration-rapidly-changing-world/ [Accessed: July 29, 2019]

[73] Radjou N. Polycentric Innovation: The New Global Innovation Agenda for MNCs [Internet]. Harvard Business Review. 2009. Available from: https://hbr.org/2009/11/polycentric-innovation-the-new.html

[74] Nyangon J, Byrne J. Diversifying electricity customer choice: Reving up the new york energy vision for polycentric innovation. In: Tsvetkov P, editor. Energy Systems and Environment. London: IntechOpen; 2018. pp. 3-24

[75] Liu Z, Liu S. Polycentric development and the role of urban polycentric planning in China's mega cities: An examination of Beijing's metropolitan area. Sustainability. 2018;**10**(5):1588

Chapter 7

Establishing Property Rights and Private Ownership: The Solution to Malinvestment in the Energy Sector in Developing Countries

Tam Kemabonta

Abstract

There are over 800 million people in the world without access to modern forms of energy services, like electricity, cooking gas, and LPG. This has been called energy poverty. Most studies in the field of energy poverty address the issue from an absence of technological or financial resources perspective. They address the problem as energy in itself having an objective inherent value, more or less addressing the symptoms of the problem and not the problem itself. In this chapter, a new paradigm that addresses the problem of energy poverty and malinvestment is introduced. This paradigm, utilizing the theory of economic calculation and the use and exchange value embodied in the subjective value theory, makes a case for the importance of private property rights in the factors or means of production for modern forms or energy such as electricity. The Nigerian energy sector is used as a case study for this.

Keywords: natural energy resources, energy poverty, rural electrification, economic calculation, subjective theory of value, property rights

1. Introduction

According to the International Energy Agency (IEA) there are over 840 million people worldwide without access to electricity [1]. In sub-Saharan Africa (SSA), the electrification rate is 47% while in Nigeria it is about 55, and 39% for those that live in rural communities. Hence there are over 95 million people without access to electricity in Nigeria [2]. Due to the high cost of cost of grid extension to these people, providing electricity, especially to those in remote communities can be prohibitive.

There is also the problem of reliability for those with a connection to the grid. This has led to an increase in the cost of doing business and, according to a 2016 International Monetary Fund (IMF) report was, in part, a factor in the decline on the country's economic growth [3]. Nigerian firms connected to the grid experience about 32.8 outages per month and the average duration for an outage is 8 hours [4]. Power outages have been estimated to cost the Nigerian economy over $7 billion, equivalent to 2.26% of the country's GDP and about 56.9% of its 2015 national budget [5].

In 2015, the United Nations General Assembly adopted a set of 17 goals to bring shared prosperity and peace to the nations of the world by 2030 [6]. These *2030 Agenda for Sustainable Development* goals are generally known as the UN SDGs. While most of the SDGs build off one another, the seventh, which to ensure access to affordable and clean energy and the thirteenth, which is climate action are considered complementary. Today, many governments, multilateral organizations and international NGOs are addressing the problem of energy poverty with climate change mitigation and adaptation policies [7, 8]. A sort of "kill two birds with one stone" is a strategy.

Many studies have shown that the access and utilization of modern forms of energy—electricity—is directly proportional to economic development. In general, people with better access to electricity have better standards of living [9–11]. The development of China and India in the last 3 decades lends credence to this. But these countries increased their electrification rates, through state-owned utility entities which utilized technologies that made them top emitters of greenhouse gases [12]. This is barely the sustainable economic development the UN had in mind.

While all these studies and strategies for addressing the problem of energy poverty in a sustainable manner that does not add to the problem of climate change, are important, they only deal with the symptoms and not the problem.

2. The current paradigm of energy poverty research

The current paradigm is to first ask, why do people in the developing world experience energy poverty? The answer is usually, they have no access to capital, hence; there is a need to transfer capital in the form of foreign aid or innovative financing instruments like green bonds and other thematic bonds, guarantees, pension funds, microfinance investment funds, and other institutional investment to these countries to addresses the problem of energy poverty. **Figure 1** shows the current paradigm of energy poverty research.

A more effective question that attempts to get to the root of the problem was suggested by Swedish economist, Per Bylund:

> *What causes poverty? Nothing. It's the original state, the default and starting point. The real question is, what causes prosperity?*

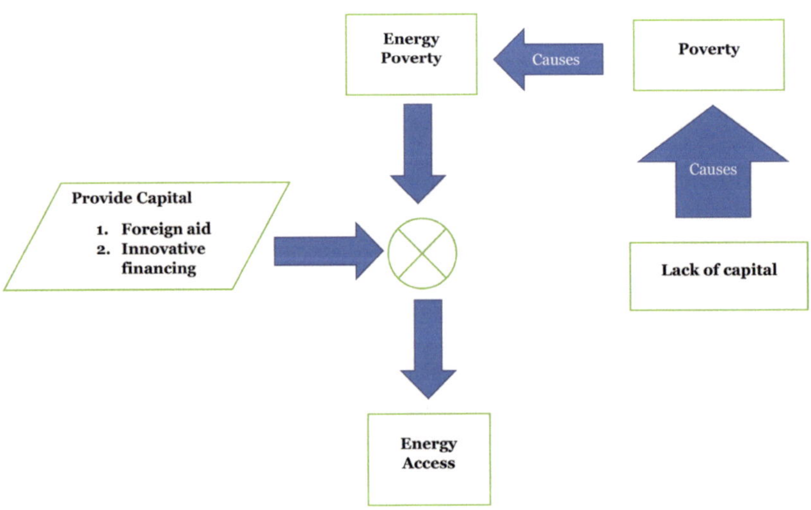

Figure 1.
The current energy poverty research paradigm.

DOI: http://dx.doi.org/10.5772/intechopen.91039

For the sake of our discussion in this chapter, professor Bylund's statement could be rephrased as:

> *What causes energy poverty? Nothing. It's the original state, the default and starting point. No community just out of thin air has the infrastructure that makes electricity available. The real question is, what causes electricity prosperity?*

This should be the real question and it is the question I attempt to answer in this chapter. The answer to this question will lead to a better understanding of the problem of energy poverty.

3. The new paradigm of energy research

Many analysis and studies on energy poverty have addressed it from the position that energy itself has an intrinsic value, an "objective value." One prominent researcher has called it the only universal currency [13]. But in an economy, people use the economic goods, time and labor they own to make ends meet. That is, they take "action" with these goods in the direction they believe will bring them the most satisfaction and prosperity. These economic goods also include the factors or means of production—land, labor and capital. To effectively utilize these goods to make ends meet they need to value the different goods they have, and the ends they want met, eventually creating a scale of value. It is the ends that are at the top on their scale of values, under the conditions and circumstances at the time the decision is made, that they will be inclined to utilize their economic goods to satisfy [14]. All goods or ends on a person's scale of value has a "use-value" which is intrinsic only to that person.

For people to use the means (economic goods) they own in the direction of the ends they want met they must exchange it for the ends they want met. This end could be a consumer or capital good. Hence, the means they own also acquires an exchange value [14]. This exchange value for any commodity e.g. electricity, is derived from the subjective value that is identified from the other commodity or a certain amount of that commodity that one plans to exchange the electricity for or vice versa.

Professor Thomas Taylor explains this thus:

> *... any particular good takes on both a use value and an exchange value. Each of these values reflects the satisfaction that can be expected to come by way of employing the good; the good can be employed either for direct use or as a means of obtaining some other good through outright exchange with another person. The controlling valuation for decision and action is always the greater of the two alternative satisfactions. If the good's use value exceeds its exchange value, the good will be put to direct use or held for eventual direct use, and its exchange value will be forgone. On the other hand, if its exchange value exceeds its use value, the good will be utilized for exchange purposes or held for possible exchange at some time in the future [15].*

In an advanced economy, where specialization and division of labor are the order of the day, people produce goods essentially for exchange, goods acquire an exchange value in addition to the use-value. Because many producers have no plan to utilize the goods they produce but essentially to exchange them for money, the use-value of those products to the producer is zero, but that good has to have an exchange value for the producer to sell it.

Modern forms of energy—like electricity—are an end. And for it to be met it must be up in position on a person's scale of value and must have an exchange value, by which such a person can use to make their decision at that specific time the decision is to be made. Electricity does not have an objective intrinsic value. It has a subjective use and exchange value. For example, a person can decide to make a down payment for a car and not for an electricity bill for this month, and another person may decide to pay for electricity and not buy the car. Both individuals—whom in this case are consumers—have different use-values for electricity.

4. Economic calculation, discovery, coordination and incentives

People value the same commodity differently. Even the same person will value a certain commodity differently at different times, hence such a person will allocate their resources differently in other to acquire the commodity the value. In an advanced economy characterized by specialization, division of labor and knowledge, multiple steps of production in producing consumer goods, multiple economic goods that can be used to produce one commodity or an alternative one and different people with different scales of values, of varying degrees of need, competing for scarce resources which have alternatives uses, there has to be a way of deciding what the most efficient use of resources, that they are utilized to satisfy the needs of people with the most intensity before satisfying those of lower intensity. There must be a lowest common denominator by which this objective calculation can be made. This common denominator which is reflected in prices that are derived from the exchange value of the different economic goods under consideration is money. Money is not a measurement of value and prices are not measured in money instead they are amounts of money [16]. Hence to determine the most efficient allocation of resources money prices are the most efficient method for calculation. This is called theory of economic or monetary calculation, introduced by economist Ludwig von Mises in 1920 and expanded later in the 1930s upon by Nobel prize winning economist Friedrich Hayek [16, 17].

Consider two examples:

The first one from economist Leland Yeager:

Consider the issues of providing public transportation in a city.

> *Should it be supplied by busses burning gasoline, by electric streetcars, in some different way, or not at all? The economically efficient answer depends on more than technology and the physical availability of inputs. It depends also on substitutabilities and complementarities among inputs, on alternative uses of those inputs, and on consumers' subjective appraisals of various amounts of the various outputs of those alternative uses, as well as on appraisals of various amounts of various kinds of public and private transportation. The economically efficient answer even to the relatively simply question of local transportation depends, in short, on unimaginably wide ranges of information conveyed, in abbreviated form, by prices [18].*

The second from Mises:

> *The art of engineering can establish how a bridge must be built in order to span a river at a given point and to carry definite loads. But it cannot answer the question*

> *whether or not the construction of such a bridge would withdraw material factors of production and labor from an employment in which they could satisfy needs more urgently felt. It cannot tell whether or not the bridge should be built at all, where it should be built, what capacity for bearing burdens it should have, and which of the many possibilities for its construction should be chosen [16].*

Money prices allow for the calculation which tells us what the most appropriate allocation scarce resources should be. It also allows for the necessary discovery and coordination needed for decision making with respect to the allocation of scarce resources in areas of the economy where they are most needed.

In today's economy, we know that electricity is demanded by different customers but to varying degrees. All these customers value electricity differently. A single person living in a small apartment in Lagos, Nigeria, values electricity differently from a Datacenter in Kaduna, Nigeria. All these entities, the single person and the Datacenter, create their scale of values based on the use value of electricity. Then they decide how much they value the money they have to exchange for electricity. A combination of all these valuations in the market determines what the exchange value of electricity will be, which is reflected in monetary prices.

Now electricity producers can calculate if the use-values of the factors of production they have at their expense through different combinations to produce electricity is less than the electricity exchange value reflected by the monetary price of electricity on the market. If it is, the producer then allocates the scarce resources accordingly to produce electricity. How the producers combine the factors of production at their expense to produce electricity becomes important. The producer may well decide that a diesel generator is cheaper than a solar panel using the monetary price of electricity as a guide. It is important to note that these monetary prices can change and are always changing because people have different use-values for different things at different times. Producers must always be cognizant with the monetary prices to decide the most effective way to allocate resources for electricity production. Profits or losses are a way of ensuring that the producers are making accurate calculations and resource allocations. If producers make a profit then they are allocating scarce resources efficiently, if they make a loss they must stop and change their course of action. If they do not, because they necessarily have limited resources, they will go bankrupt and stop wasting scarce economic resources. And everyone in the economy is better for it.

For all these to be possible, they must be a market in the factors of production—land, labor and capital. For there to be market in the factors of production there must be private ownership in the factors of production. As Mises puts it "... in the absence of market prices for the actors of production, a computation of profit or loss is not feasible" [14]. When computation of profit and loss is not feasible producers are groping in the dark, and this leads to either wastages or shortages of economic resources. Mises points this out:

> *Under a system based upon private ownership in the means of production, the scale of values is the outcome of the actions of every independent member of society. Everyone plays a two-fold part in its establishment first as a consumer, secondly as producer. As consumer, he establishes the valuation of goods ready for consumption. As producer, he guides production-goods into those uses in which they yield the highest product. In this way all goods of higher orders also are graded in the way appropriate to them under the existing conditions of production and the demands of society. The interplay of these two processes ensures that the economic principle is observed in both consumption and production. And, in this way, arises the exactly graded system of prices which enables everyone to frame his demand on economic lines [19].*

> *Monetary calculation is the guiding star of action under the social system of division of labor. It is the compass of the man barking upon production. He calculates in order to distinguish the remunerative lines of production from the unprofitable ones, those of which the sovereign consumers are likely to approve from those of which they are likely to disapprove. Every single step of entrepreneurial activities is subject to scrutiny by monetary calculation. The premeditation of planned action becomes commercial pre-calculation of expected costs and expected proceeds. The retrospective establishment of the outcome of past action becomes accounting of profit and loss [14].*

Deploying renewable energy technologies such as distributed solar photovoltaics (PV) panels and wind turbines have been proposed as solutions to the problem of energy poverty in many developing countries, especially those who live in rural communities. These communities are usually identified as underserved or unserved [20]. The use of solar PV technologies has increased over the last two decades. This is, in part, due to the decrease in the cost renewable energy technologies and creative financial structures that enable them to become affordable [7, 8, 20, 21]. This way, developing countries can have access to electricity and keep down the likelihood of contributing to climate change. In one fell swoop, we can meet the UN SDG 7 and 13 [6]. But as we will see it is not that simple. Looking at the problem in terms of technology or a lack thereof does not get to the crux of it. North Korea has a nuclear weapons program but still has most of its people living in poverty, with 1 in 5 of its population not having access to clean water or adequate sanitation [22]. The old Soviet Union was the first country to put a man in space but millions of its people were dying of hunger, and the nation was replete with the wastages and shortages of different goods because of the lack of private ownership in the means or factors of production [23]. Without the private ownership in the means of production, in this case, land and fossil fuels, malinvestments in the energy sector will occur, causing wastage and shortages, which will neither achieve the goals of SDG 7 nor 13.

This is because those who live in the rural communities in the developing countries that are usually targets of renewable energy-based electrification projects do not have secured property rights in the land they live on. Therefore, they cannot effectively place a value on the land and its potential use for development of electrification projects with respect to the opportunity cost of the other things on their scales of values. This makes deriving monetary prices for electricity difficult.

Governments in these countries, through their electricity regulatory agencies or state-owned energy utilities (gas and electricity) institute price controls on electricity, which defeats the calculation, discovery, coordination and incentive process of the price mechanism, leading to either shortages in the production of electricity or wastages through overproduction of electricity. A 2016 World Bank report showed that most utilities of 39 countries in sub-Saharan Africa are almost always insolvent [24].

The study goes on to say:

> *Of the 39 countries studied, only the Seychelles and Uganda were fully recovering their operational and capital costs. In only 19 countries did the cash collected by utilities cover operational costs; just 4 of these countries were also covering half or more of capital costs, based on new replacement values of current assets. Such large funding gaps prevent power sectors from delivering reliable electricity to existing customers, let alone expand supply to new consumers at an optimal pace [24].*

Twenty-one out of 48 sub-Saharan countries have no private participation in their electricity sectors. They still have state-owned vertically integrated utilities. Most of the others have different degrees of private ownership and participation in the distribution, transmission and generation electricity [25].

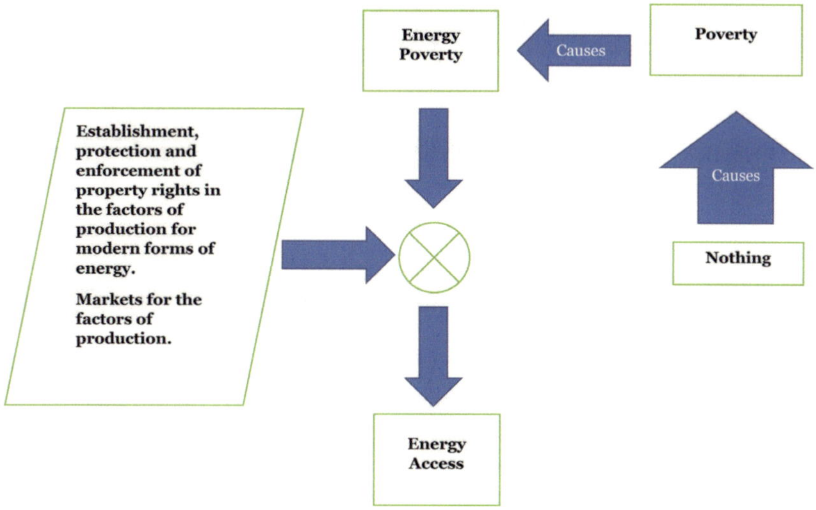

Figure 2.
The new energy poverty research paradigm.

In almost all cases, government agencies have controls on the prices of electricity, which is almost always not cost reflective [20, 24, 25]. The lack of private ownership in the factors of production for electricity is the major reason why many SSA countries still experience energy poverty, especially in the form of electricity. Hence, how energy poverty research is conducted needs to change. **Figure 2** shows what the new paradigm of energy poverty research should be.

5. Energy poverty in Nigeria and sub-Saharan Africa

To solve malinvestments in the energy sector and the problem of energy poverty there must be a private ownership of the factors of production—land, labor and capital—used to make modern forms of energy services available. When this not the case, appropriate use-values and exchange values cannot be arrived at, because you can only effectively value what you own, to decide if it is worth it to exchange for what you want or not. Hence, there is no market for the factors of production used to make electricity. When there is no market, there are no market prices. Without market prices or attempting to fix a market prices, like SSA governments do, will always lead to a wastage or shortage of electricity production.

The major factors of production for electricity under consideration are land and its tributaries, such as mineral and natural energy resources like oil, natural gas, iron ore, etc. If any of these factors of production cannot be privately owned, you cannot have an efficient allocation of resources to provide electricity in the quality and quantity demanded.

It does not matter if a country is resource rich, as long as private ownership does not exist in those resources and there are no market prices for the resources it will be wasted at the expense of the entire economy. Let us consider the example of Nigeria's energy sector.

Nigeria has enough natural gas and oil resources to solve its energy problem. It is the largest oil producer in Africa and the fourth largest exporter of liquified natural gas (LNG) [26]. Oil and natural gas are factors of production in the production of electricity. The land needed to build the infrastructure (capital goods), such as pipeline, generating plants, transmission and distribution lines, which will enable converting oil and natural gas to electricity is a factor of production. In Nigeria private property rights in land is shaky and non-existent in natural resources [27–29]. Hence, the economic/monetary calculation is difficult, leading to a lack of market prices in electricity, which results in a shortage of every consumer good—such as, electricity and petrol—that oil and natural gas is used to produce. Nigerians are consistently faced with extensive petrol shortages and chronic electricity outages [20, 26, 30].

6. Brief overview of the Nigerian power sector

In 2005, the Nigerian Electric Power Sector Reform Act (EPSRA) became law. This required a breakup of the state-owned vertically integrated electricity and for the allowance of the private sector into the business of producing and distributing the commodity—electricity [30, 31]. By 2013, the unbundling and eventual sale of the state-owned utility's—Power Holding Company of Nigeria (PHCN)—assets was complete. This led to 11 distribution companies (DISCOs), with exclusive rights to geographic service territories, six generating companies (GENCOs) and one transmission company, which is called the Transmission Company of Nigeria (TCN) [30]. The GENCOs where completely sold off to private investors, the government retained a 40% stake in the 11 DISCOs and 100% government ownership was maintained for TCN. A Bulk trader, the Nigerian Bulk Electricity Trading (NBET), was created to buy electricity from the GENCOs (both the original ones and the newer Independent Power Plants (IPPs)) and then sell to the DISCOs. The DISCOs could only buy from NBET and not directly from the GENCOs/IPPs. Finally, a regulator was created. This is the Nigerian Electricity Regulatory Commission (NERC) [30–33]. Refer to **Figure 3** for a graphical representation of the Nigerian electricity sector.

The sector is still replete with government intervention and ownership. The DISCOs continuously experience capacity and insolvency issues [30]. With all the risks of electricity theft, old infrastructure and vandalism NERC still does not allow the DISCOs charge cost effective tariffs [30, 34]. The problems of the Nigerian energy sector, like that of many other developing countries, are many and well documented in the old paradigm of energy poverty research [4, 20, 24, 30, 32–36]. But as we can see there has been little to no amelioration in the last 3 decades.

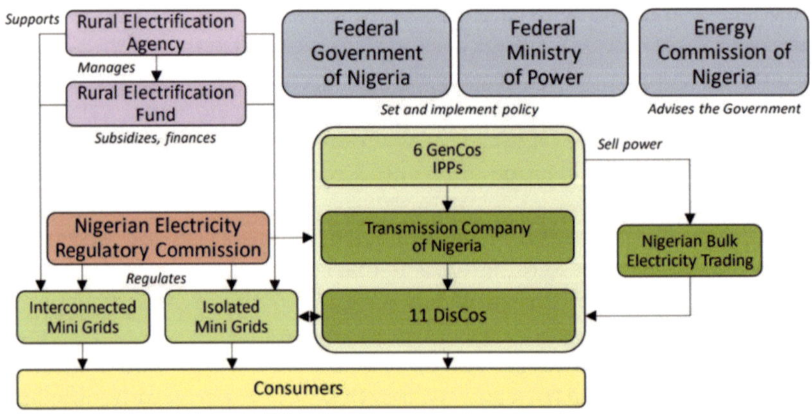

Figure 3.
Overview of the Nigerian electricity sector [32].

The reason why these problems have proven protracted is simple: no private property in electricity factors of production, hence no market prices, hence no economic calculation, hence wastage or shortages, exacerbating energy malinvestment and energy poverty. When in an economy there is very little private ownership in any sector, consumer goods from that sector do not get produced in the quantity or quality they are demanded. This is the new paradigm of energy poverty research introduced here.

7. Private ownership in the factors of production for electricity—land and natural energy resources

In this section we will look at how lack of private property in two important factors of production for the electricity—land and natural energy resources—has caused energy poverty in Nigeria.

All land in Nigeria, apart from those vested in the federal government and its agencies is under the control of the state governors. The lands are held in "trust and administered for the use and common benefit of all Nigerians..." in urban areas by the state governors and in rural areas they are managed by the local governments [27]. Nigerians can only get access to the land through a certificate of occupancy (C of O), which is valid for 99 years, given by the state governor or on the governor's behalf. The state governor is also responsible for land use decisions. For example, if somewhere is to be used as a residential, commercial or industrial area. Hence, land can also be taken from anyone with a C of O if the government decides the land should be used for something else. Private property rights to land in Nigeria is shaky at best [28]. There have been numerous violent forced evictions of people by the government from lands that have been in their families for decades, because a governor decided to designate that piece of land for development. Sometimes these governors completely ignore court injunctions attempting to stop them from carrying out their violent decimations of affected communities [28, 37, 38].

If private property to land in Nigeria is on shaky foundations, private property to natural energy resources is nonexistent. All natural resources within the geographic domain of Nigeria are owned by the federal government.

> *...the entire property in and control of all minerals, mineral oils and natural gas in under or upon any land in Nigeria or in, under or upon the territorial waters and the Exclusive Economic Zone of Nigeria shall vest in the Government of the Federation and shall be managed in such manner as may be prescribed by the National Assembly [29].*

Therefore, even if a person had a C of O to a piece of land and oil was discovered on that land, immediately the government can expropriate that land from such a person. Public ownership of natural energy resources has wreaked havoc on the country, bringing about poverty, environmental degradation and mismanagement of windfall profits. This is because when there is no private ownership, hence no market prices in natural energy resource and the mechanism for economic calculation, coordination, discovery and incentives breaks down, leading to shortages or wastages. This has been the case of the Nigerian energy sector.

The government does not only own and control the exploitation of all natural energy resources, it also controls the prices of the consumer products produced from these resources:

> *The Minister may by order published in the Federal Gazette fix the prices at which petroleum products or any particular class or classes thereof may be sold in Nigeria or in any particular part or parts thereof [29].*

The Nigerian government benefits from its large natural energy resources reserves by licensing and forming joint ventures with International Oil Companies (IOCs) and Local Oil Companies (LOCs). These companies exploit the resources with reckless abandon leaving environmental degradation in their wake leaving many of their host communities unlivable. Over 2 million barrels of oil were spilled between 1976 and 1996 [39]. Since the IOCs and LOCs do not own the natural energy resources, they are mostly interested in the short term profits of exploiting the resources, having no incentive to preserve the long term capital value of the resources as a private owner would, they flare natural gas, a useful byproduct of the oil drilling process, further polluting the environment [40, 41]. This has led to violent conflicts between the IOCs/LOCs and the communities. The Niger Delta, where most of Nigeria's natural energy reserves are found is one of the least developed and poorest regions of the country [41, 42].

Max Siollun, a historian, in his book, *Oil, Politics and Violence: Nigeria's Military Coup Culture (1966–1976),* presented the quintessential example of how the Nigerian government has tried to manage the countries natural energy resources for the betterment of its economy but instead has repeatedly failed wreaking havoc on the economy:

> *The influx of petrodollars into government coffers also amplified both the Nigerian government and people's developmental ambition.... The (Federal Military Government) FMG proved ineffective at managing the wealth, and was unable to use it to significantly increase Nigerians' living standards. Although the oil boom created a tiny coterie of powerful economic oligarchs and patronage system amongst senior military officers, their families and their civilian associates, living conditions for the rest of the population either remained stagnant or deteriorated. This created the paradox of a rich country with poor people. Gowon (the Head of State) described the problem as "want in the midst of plenty" and observed that Nigeria's problem was not lack of money, but how to effectively spend its sudden new found wealth.*
>
> *Civil perceptions that Nigeria was "rich" also made the population impatient for the oil boom wealth to trickle down to the society at large. In an attempt to distribute federal wealth to workers, the FMG in January 1975 decided to award public sector employees massive pay rises exceeding 100%...*
>
> *The increased spending power of public sector workers led traders to increase their prices, fueling inflation and wiping out the economic benefits the pay rises were intended to create. Private sector workers then went on strike to demand pay rises for themselves [43].*

A lot has been written about the resource curse, first introduced by economist Richard Auty and popularized by economists Sachs and Warner [44, 45]. The resource curse theory is an attempt to explain why some countries with abundant natural resources are usually worse off economically or have the least level of economic development, compared to prosperous countries with little or no natural resources [45]. What many of these studies fail to take into consideration is that in places where private property in land and natural resources exist, are secured and can be easily exchanged from one party to another, the resource curse was never an issue.

For example, let us look at the difference between the issue of oil drilling in the Arctic National Wildlife Refuge (ANWR) and the Audubon Society's Paul J. Rainey wildlife sanctuary (PRWS) in Louisiana—both places in the United States. The ANWR is owned by the government while PRWS is owned by a private organization—the Audubon Society. Policy analyst Fred Smith described the situation thus:

> *Both of these areas are valued by environmentalists. Both also sit above oil deposits. In the case of ANWR, we have witnessed political gridlock. To put it very simply: the environmentalists want it preserved, and the oil companies want to drill. ANWR is a political football in the Congressional debates over environmental and energy policy. Rainey is different. This refuge is owned privately by the Audubon Society, rather than by the federal government. At this site, Audubon has the ability to exclude all visitors and activities that could damage the refuge or threaten the animals that live and breed there. Audubon could have prevented all oil development at Rainey. They chose not to do so. Preventing oil development would have required foregoing the economic benefits of that development-economic benefits that could fund other environmental efforts. As a private owner, Audubon had an incentive to reconcile the very same interests that are in conflict in the case of ANWR. Audubon developed an oil extraction plan that would allow drilling but also protect Rainey's ecological values. They did so by making accommodations: no drilling during the breeding season; a smaller oil platform; spill prevention and containment plans to prevent contamination, and the like. Oil production has been occurring under these conditions at Rainey for over twenty years with little problem. Because of Audubon's private ownership, it was possible to integrate the human economic and ecologic concerns. Private ownership encouraged people to work toward this type of win-win solution. Politics too often encourages conflict and a zero-sum game. Where politics has been dominant-as in the case of ANWR---conflict, not accommodation, has been the rule [46].*

Why would an organization like the Audubon Society allow oil development on the PRWS which is "home for deer, armadillo, muskrat, otter, mink and more than 50,000 snow geese... also is the site of a number of oil and gas wells, and provides grazing land for private cattle herds" [47]. This is because the PRWS is private property, the natural energy resources in the land, is a factor of production and can be sold on the market. With these market prices, Audubon can effectively calculate economically what the cost and benefits of allowing oil development on the PRWS would be to them. This would include what allowing this could do to its reputation, since Audubon is a natural conservancy organization. Audubon can decide subjectively what the use-value of that PRWS is, and since there is a market for the natural energy resource, it can achieve an exchange value. Audubon discovered that the exchange value was more than the use-value to them, and this presented a platform for them to decide. Audubon and the oil company were both better off as a result of the transaction. Audubon, throughout the lifetime of the contract with the oil company made over $20 million in royalty checks. The oil company, as long as they met the Audubon's conditions, like ensuring the environmental integrity of the PRWS ecosystem, could exploit the natural energy resource [47, 48].

This is only possible when natural energy resources are private owned. Nigeria's natural energy resources cannot lead to economic development until private property in the resources themselves exist.

8. The mini grid industry in Nigeria: a possibility for moving forward

While in Nigeria land ownership in urban areas is largely contested due to the higher value on them, in the rural communities this is not always the case. The laws that govern land use in the country are largely ignored in those places. Okafor et al. stated that while "the land use Act provides that 'all lands in rural areas, be under the control and management of the Local Government, within the area of jurisdiction of which the land [is] situated,' which implies that there will be no more open market transaction, yet this is still in practice in the area." [28].

Around 2013 entrepreneurs went into rural communities to develop and deploy small power systems in these communities providing them electricity. These rural communities are usually off grid. In these cases, the community members were able to place a use-value on the land they had at their disposal and compare it with the exchange value of electricity. In many cases the developers were allowed to install their solar + diesel or storage systems in the communities [20, 32].

What is important to note is the tariffs the mini grid electricity companies charged the community members. Their tariffs were between $0.38 and $0.51. This is more than the tariff the DISCOs charge on the main grid, which is usually between $0.064 and $0.080 for the Eko Distribution Company [30, 49]. It may seem like the off-grid rural community members are paying more, but this is not the case. In some of these communities they utilize small petrol generators to provide for their electricity needs. For many of them to refill their generators, they have to travel many kilometers to the nearest gas station, they also have to operate and maintain their generators. These add to costs that are either the same or higher than the tariffs offered by the mini grid electricity companies. Hence making what may look like a high tariff to third parties, a perfect price, they are willing to pay, and the mini grid company is willing to accept for electricity.

In 2016 the Nigerian government, through NERC released a regulatory framework for the development of renewable energy based mini grids [50]. This was done to help the nation accelerate its electrification plans. Nigeria plans to achieve universal energy access by 2030 [4]. The policy was also instituted to regulate the mini grids that had started springing up in different off-grid communities before 2016. It was also meant to guide the new mini grid industry that has been estimated to be $8 billion-dollar industry annually [51].

Mini grids below the size of 100 kW do not need to apply for a permit, those below the size of 1 MW need to apply for a permit and those above 1 MW are treated as IPPs and must get a license. The mini grid companies are also able to determine the tariffs they charge the community members. Hence, they are free to negotiate with the community to determine what price they are willing to pay and if this price can economically justify deploying and operating a mini grid electrification asset in the community [32]. Since the policy was passed, the Rural Electrification Agency (REA), the government agency responsible for providing technical and financial assistances to rural electrification project has become very involved in the industry [32, 51]. Many rural communities are now receiving government oversight, the danger of this is that property rights in land and other factors of production like solar panels and batteries, can become contested soon, leading to government expropriation of the land in these communities, in the name of using them for the development of electricity projects, and if this happens the rural communities will not achieve the energy access, they will become worse off and as evidence shows, will probably become like the impoverished Niger Delta communities.

DOI: http://dx.doi.org/10.5772/intechopen.91039

9. Conclusion

To achieve universal electrification or the UN SDG 7 by 2030 in Nigeria or other developing countries, there needs to be a paradigm shift in how energy poverty and malinvestment in the energy sector research is conducted. In this chapter a new paradigm that takes into consideration the importance of private ownership of the factors or means of production for modern forms of energy services like electricity and the use and exchange values of these factors or means founded on the theory of economic calculation and the subjective theory of value is introduced. More studies on the applications of the economic theory of calculation and the subjective theory of value in energy development in developed and developing countries are needed and important. It is my hope that this chapter serves as a catalyst for this research.

Author details

Tam Kemabonta
Independent Scholar

*Address all correspondence to: tamkemabonta@energypoliticking.com

IntechOpen

© 2020 The Author(s). Licensee IntechOpen. Distributed under the terms of the Creative Commons Attribution - NonCommercial 4.0 License (https://creativecommons.org/licenses/by-nc/4.0/), which permits use, distribution and reproduction for non-commercial purposes, provided the original is properly cited.

References

[1] IEA, IRENA, UNSD, WB, WHO. Tracking SDG 7: The Energy Progress Report 2019. Washington DC; 2019

[2] The World Bank Nigeria Electrification Project (P161885), International Development Association Project Appraisal Document on a Proposed Credit in the Amount of SDR 243.4 Million (Us$350 Million Equivalent), Report No: PAD2524

[3] International Monetary Fund (IMF) World Economic Outlook; 2016

[4] World Bank Program-For-Results-Financing, Power Sector Recovery Program, Final Report, Environment and Social Systems; 2018

[5] Amadi HN, Okafor ENC, Izuegbunam FI. Assessment of impact of outages in selected electricity intensive industries in Nigeria. IMPACT: International Journal of Research in Engineering & Technology. 2016;**4**(10):9-24. ISSN(P): 2347-4599; ISSN(E): 2321-8843

[6] United Nations General Assembly. Transforming our world: The 2030 Agenda for Sustainable Development; 2015

[7] Zerriffi H, Wilson E. Leapfrogging over development? Promoting rural renewables for climate change mitigation. Energy Policy. 2010;**3**(4):1689-1700

[8] Levin T, Thomas VM. Can developing countries leapfrog the centralized electrification paradigm? Energy for Sustainable Development. 2016;**31**:97-107

[9] European Commission. Science for environment policy: Relationships between energy consumption and economic growth investigated; 2013

[10] Wang S, Li Q, Fang C, Zhou C. The relationship between economic growth, energy consumption, and CO_2 emissions: Empirical evidence from China. Science of the Total Environment. 2016;**542**:360-371

[11] Ping WL. Study on the relationship between economic development and energy consumption in Henan Province of China. In: 2008 International Seminar on Business and Information Management, Wuhan; 2008. pp. 390-393

[12] Janssens-Maenhout G, Crippa M, Guizzardi D, Muntean M, Schaaf E, Olivier JGJ, et al. Fossil CO_2 and GHG emissions of all world countries, EUR 28766 EN, Publications Office of the European Union, Luxembourg; 2017. ISBN 978-92-79-73207-2. DOI: 10.2760/709792, JRC107877

[13] Smil V. Energy and Civilization. Boston, Massachusetts: MIT Press; 2017

[14] von Mises L. Human Action: A Treatise on Economics. Auburn, Alabama: The Ludwig von Mises Institute; 1998. pp. 201-232. The Scholar's edition

[15] Taylor TC. An Introduction to Austrian Economics. Auburn, Alabama: The Ludwig von Mises Institute; 1980. Copyright by Cato Institute

[16] von Mises L. Economic Calculation in the Socialist Commonwealth. Auburn, Alabama: The Ludwig von Mises Institute; 1920. p. 1990

[17] Friedrich H. "The nature and history of the problem"; "The present state of the debate." Collectivist Economic Planning; 1935. pp. 1-40, 201-243

[18] Yeager L. Mises and Hayek on calculation and knowledge. The Review of Austrian Economics. 1994;7(2):93-109

[19] von Mises L. Socialism: An Economic and Sociological Analysis. New Haven, Connecticut: Yale University Press; 1951. pp. 113-122

[20] Kemabonta T, Geoffrey A, Apena O, Uzzle S. What went wrong and how can we fix it: Renewable energy and mini-grid policies in Nigeria. In: IEEE PowerAfrica Conference, Abuja, Nigeria; 2019

[21] Zerriffi H. Rural Electrification: Strategies for Distributed Generation. London: Springer Science and Business Media; 2010

[22] United Nations. DPR Korea, Needs and Priorities; 2017

[23] Shmelev N, Popov V. The Turning Point: Revitalizing the Soviet Economy. New York: Doubleday; 1989

[24] Kojima M, Trimble C. Making Power Affordable for Africa and Viable for its Utilities. Washington DC: The World Bank Group; 2016

[25] Eberhard A, Gratwick K, Morella E, Antmann P. Independent Power Projects in Sub-Saharan Africa Lessons from Five Key Countries. Washington DC: World Bank Group; 2016

[26] U.S. Energy Information Administration. Country Analysis Brief: Nigeria; 2016

[27] Land Use Act. Chapter 202, Laws of the Federation of Nigeria; 1990

[28] Okafor BN, Nwike EC. Effects of the land use act of 1978 on rural land development in Nigeria: A case study of Nnobi. British Journal of Environmental Sciences. 2016;**4**(3):1-16

[29] Constitution of the Federal Republic of Nigeria 1999, Section 44(3) and the Nigerian 2004 Petroleum Act, Section 6(1)

[30] Kemabonta T, Kabalan M. Using what you have, to get what you want – a different approach to electricity market design for local distribution companies (DISCOs) in Nigeria. In: 2018 Global Humanitarian Technology Conference (GHTC), USA

[31] Federal Republic of Nigeria. Electric Power Sector Reform Act; 2005

[32] World Bank. Climate Investment Funds, ESMAP. Mini grids in Nigeria: A case study of a promising market; 2017

[33] Usman ZG, Abbasoglu S. An overview of power sector laws, policies and reforms in Nigeria. Asian Transactions on Engineering (ATE ISSN: 2221-4267) Volume 04 Issue 02; 2014

[34] Gatugel Usman Z, Abbasoglu S, Tekbiyik Ersoy N, Fahrioglu M. Transforming the Nigerian power sector for sustainable development. Energy Policy. 2015;**87**:429-437

[35] Joseph IOO. Issues and challenges in the privatized power sector in Nigeria. Journal of Sustainable Development Studies. 2014;**6**(1):161-174

[36] Kalu U, Agbaeze EK. Deregulation of the Nigerian power sector on performance: A review. European Journal of Scientific Research. 2018;**148**(3):377-385

[37] Agbola T, Jinadu AM. Forced eviction and forced relocation in Nigeria: The experience of those evicted from Maroko in 1990. Environment and Urbanization. 1997;**9**(2)

[38] Olutimehin Adegbeye. Homeless by design. Latterly, medium.com; 2017

[39] Ajide OM, Isaac OO. An assessment of physical impact of oil spillage using GIS and remote sensing technologies: Empirical evidence from Jesse Town, Delta State, Nigeria. British Journal of Arts and Social Sciences. 2013;**12**(2):235

[40] Rothbard M. Making Economic Sense, Government Vs. Natural Resources. Auburn, Alabama: Ludwig von Mises Institute; 2006

[41] Udok U, Akpan EB. Gas flaring in Nigeria: Problems and prospects. Global Journal of Politics and Law Research. 2017;**5**(1):16-28

[42] Ukaga O, Ukiwo UO, Ibaba IS (edited). Natural Resources, Conflict, and Sustainable Development in the Niger Delta. Natural Resources, Conflict and Sustainable Development: Lessons from the Niger Delta. New York: Routledge, Taylor & Francis Group; 2012

[43] Siollun M. Oil, Politics and Violence: Nigeria's Military Coup Culture (1966-1976). New York: Agora Publishing; 2009

[44] Auty R. Sustaining Development in Mineral Economies: The Resource Curse Thesis. London and New York: Routledge; 1993

[45] Sachs JD, Warner AM. Natural Resource Abundance and Economic Growth. National Bureau of Economic Research, Working Paper 5398; 1995

[46] Smith FL Jr. Sustainable development – A free market perspective. In: Boston College Environmental Affairs Law Review. Article 8. 1993;**21**(2)

[47] Baden J. Destroying the Environment: Government Mismanagement of Our Natural Resources. National Center for Policy Analysis. Policy Report No. 124; 1986. p. 34

[48] Baetjer H Jr. Economics and Free Markets: An Introduction. Washington DC: Cato Institute; 2017. pp. 155-156

[49] Nigerian Electricity Regulatory Commission (NERC). Press Release: New Tariff Regime Removes Fixed Charges; 2015

[50] International Energy Agency (IEA). Nigerian Electricity Regulatory Commission Mini-Grid Regulation 2016; 2017

[51] The Nigerian Economic Summit Group (NESG). Minigrid Investment Report: Scaling the Nigerian Market; 2018

Section 4

Case Studies

Chapter 8

The Electrification-Appliance Uptake Gap: Assessing the Off-Grid Appliance Market in Rwanda Using the Multi-Tier Framework

Olivia Muza

Abstract

The structure of the electricity system includes universal access to electricity that is adequate, available, reliable, affordable, legal, convenient, healthy, and safe and the efficient (inefficient) use of the electricity. Quality of access also influences clean energy technologies and electrical appliance purchase, ownership, use and perceived value (uptake, hereafter). Also, improved uptake assists in closing systemic gaps between rural and urban areas and grid and off-grid communities. Rwanda is projected to attain full electrification by 2024 (inclusive of all sectors: consumptive, productive and services). In this context, the East African country has articulated support mechanisms for off-grid market players through technical assessments and siting incentives. However, studies that focus on characterising diffusion and uptake of clean energy technologies and electrical appliances in mini-grid sites (market) are crucial to understand the emerging trends in off-grid rural electrification. This chapter contributes to this emerging discourse by proposing a four-fold demand side characterisation approach which (i) conducts a systemic review of literature to identify emerging off-grid themes as they relate to the multi-tier framework (MTF) and vice-versa, (ii) uses existing data to characterise the off-grid market (based on a typical village load), (iii) demonstrates the tariff regime changes using two payment methodologies (willingness to pay (WTP) and ability to pay (ATP)) and (iv) projects the 2024–2032 consumptive energy demand (using a simplified relation between appliance, it's rating and duration of use). Results of this characterisation demonstrate global and local level (glo-cal) literature gaps meriting a localised MTF assessment. The purpose of the localised assessment reported in this Chapter was therefore to understand appliance uptake gaps at the user level. The typical village load is basic (implying low energy demand). Ceteris paribus, higher WTP and ATP by users yield higher tariffs. However, a high ATP is a business sustainability determinant than a high WTP. Because energy consumption is also dependent on how efficiently it is used by those with access, the Chapter discusses appliance efficiency as a partial definition of sustainable energy and also as an example of sustainable energy. Then, demand stimulation pathways addressing wider systemic opportunities at the intersection of the theory of change and the theory of agency and risk reduction in markets, investments and policy (derisking markets, investments and policy) are discussed. The first pathway focuses on women and youth participation in productive

IntechOpen

use activities. The second pathway highlights strategies for appliance financing such as cost-sharing and micro-credit. The final pathway considers economic activity stimulation which has multiplier effects on energy demand and consequently energy-using appliances uptake. The implications for Sustainable Citizens and markets, investments and policy innovations are contextualised in the Sustainable Energy Utility business model.

Keywords: gender, consumer behaviour, time-using appliances, time-saving appliances, off-grid households, energy access, technological innovations, consumer choice, energy efficiency, appliance efficiency, derisking innovations, markets and investments, sustainable energy utility (SEU) model, transitioning economies

1. Introduction

Off-grid energy solutions often fail due to demand side factors such as insufficient energy consumption and low uptake of energy using appliances [1]. Anticipating community energy use as development occurs and users make clean energy and electrical appliance choices is crucial for national energy planning.

For example, the multi-tier definition of electrification goes beyond access and considers the quality of energy being accessed: adequacy, availability, reliability, quality, legality, convenience, health and safety (in households, productive engagements and communities) [2]. Today, issues at the intersection of community energy use, appliance uptake, the multi-tier context and their implications for Agenda 2030 for Sustainable Development Goals (SDGs) have not received adequate attention in research literature. This is also the case for the cluster of interdependent goals addressing well-being[1] (example relevant SDGs: Goal 1 on 'Zero Poverty', Goal 5 on 'Gender Equality', Goal 6 on 'Clean Water and Energy Access', Goal 7 on 'Energy Access for All', Goal 16 on 'Peace, Justice and Strong Institutions' and Goal 17 on 'Partnerships for the Goals'). This emerging discourse contributes to energy transitions and sustainability planning in transitioning economies. The purpose of this Chapter is to provide a localised and demand side characterisation approach of diffusion and uptake of clean energy technologies and electrical appliances in mini-grid sites (market) in order to understand the emerging trends in off-grid rural electrification.

East Africa's Rwanda has articulated support mechanisms to off-grid market players, for instance, through technical assessments and siting incentives. The country is targeting 100% electrification for all its inhabitants by year 2024, of which 52% is expected from grid and 48% through off-grid connections [3]. Rwanda is also considering a variety of off-grid energy options to complement electrification targets. According to the theory of change [1], the next step after electrification is adoption of electrical appliances[2]. However, this has not always

---

[1] This is also the case with other sets of interdependent SDGs for example: those ending hunger and achieving food and nutrition security (example relevant SDGs: 2, 3, 5, 17), protecting the planet and building resilience (example relevant SDGs: 5, 12, 13, 14, 15, 17), ensuring access to sustainable energy and transport (example relevant SDGs: 5, 7, 11, 12, 17), sharing economic benefits and ensuring safety of society (example relevant SDGs: 7, 8, 10, 12, 17) and prioritising resources for local action and to accelerate implementation (example of relevant SDGs: 5, 9, 11, 17).

[2] The authors highlight that appliance uptake usually begins with lights and other typically bought household goods such as televisions, radios and mobile phones. Firms could buy machinery and refrigeration. Health centres can buy lighting or simple appliances for diagnosis and treatment. Schools may uptake appliances for evening classes.

been the case as other micro level determinants such as affordability of use [4] or the quality of service [5] influence adoption. I use the theory of agency[3] to understand the supplier-customer interface influencing uptake. Today, four factors are identified as requiring further analysis in clean energy technologies and electrical appliance uptake in Rwanda; firstly, uptake is urban centric, there is need for a rural energy use boost [7]. Electrified rural households continue to use basic appliances for lighting, phone charging and small cottage industries and humbly televisions (TVs), irons, fans, refrigerators, electric cooking stoves among others [1, 2, 7–11]. Secondly, appliance uptake differs across household classes [3, 7, 10, 12, 13]. Promotion of use across household classes is imperative. Thirdly, appliance uptake is gendered [3, 7, 8, 14]. Electric appliances that appeal to men and women may promote usage. Lastly, there are sectoral differences in energy use. It is important to forge inter-sectoral linkages for electrical appliance use.

Based on the foregoing, I argue that a localised Multi-Tier Framework (MTF) assessment is crucial to contextualise appliance insufficiency or the availability of energy inefficient appliances [9]. It underscores appliance needs in existing and newly created customer regions and relevant demand stimulation packages. It demonstrates appliances of value/benefit to improving well-being [15]. It highlights the Sustainable Citizen needs which may be more advanced than those of the customer/consumer. The possibilities of a Sustainable Energy Utility (SEU) model and policy innovations [16, 17]. The rest of this Chapter is structured as follows: in the next immediate sections. I discuss the MTF framework, global trends and local relevance. This is followed by the conceptual framework, methodology, results, discussion of findings, policy implications and conclusion.

2. The MTF

The multi-tier context has seven key check-points that can assist in carrying a localised assessment [2] (**Table 1**).

- The ability to power appliances (adequacy/capacity)

	Tier 0	Tier 1	Tier 2	Tier 3	Tier 4	Tier 5
Capacity	No electricity	1-50 W	50-500 W	500-2000 W	>2000 W	
Duration	< 4 hrs	4-8 hrs		8–16 hrs	16-22 hrs	22 hrs
Reliability	Unscheduled outages				No scheduled outages	
Quality	Low quality			Good quality		
Affordability	Not affordable		Affordable			
Legality	Not legal			Legal		
Health and Safety	Not convenient				Convenient	

Source: Ref: [2, 18].

Table 1.
MTF of household electricity.

[3] The theory of agency in marketing was underscored by early Scholars like Bergen et al., [6]. It has been applied in different contexts to explain different kinds of agency relationships. In this Chapter it focuses on the supplier-customer relationship.

- Number of day or night hours (duration)
- Reliability (electricity events and power outages)
- Whether voltage hours affect the use of desired appliances (voltage problems)
- The cost of a standard consumption (affordability- basic services less than 5% of household income)
- Whether the service is provided legally (legality)
- Absence of risk (health and safety)

In this Chapter, uptake refers to any or all of these demand side factors: acquiring (purchase, borrowing or gifting), number in built environment (ownership), how they are used (use), how often they are used (number of hours), and the benefits derived from all of these factors (user perceived value).

In the Rwandan context, the anticipated 100% electrification target coincides with its ambition of becoming an upper middle income country by 2035, a high income country by 2050 and a provider of high technology services to the wider East Africa region. If energy use will grow more quickly for households coming out of poverty than for households further up the income distribution [19] I assume that electrical appliances diffusion and energy consumption imitates an S-Curve pattern consisting of three stages of development: early; exponential and saturation [20, 21]. But in the real world energy transitions are not linear [22]. The Chapter lays out potential considerations in the off-grid sector.

3. Global trends

While electrical appliance use is a key driver of electrification rates, it is only now beginning to draw attention in developing world literature as a sustainability driver, a policy enabler and a standalone research agenda. By not using appliances, electrification and appliance related benefits: business opportunity, food security, ability to acquire knowledge, time savings and productivity among others remain elusive. Evidence from recent studies in sub-Saharan Africa include Rwanda [7], urban Ghana [23] and rural Uganda [24]. Other relevant studies cover the developing world in general but from a macro-level perspective and for a different point in time [19], others have relevance in strategic direction but are from a past era and a global north focus [25, 26]. A discussion of appliance uptake from the global south perspective has been sufficiently addressed in my previous work from a social perspective also using a Rwandan case [7]. However, from the energy consumption perspective which is the focus in this Chapter, appliance ownership is one of the key variables. In modelling domestic end use/consumption and determining load profiles using statistical or regression [27–31], engineering [32–36] and neural networks (NNs) [37–39]. It is also important to explore strategies for promoting appliance use.

4. Localising global trends

Electrification in the MTF context, decentralisation and centralisation, domestic end-use/consumption, energy as a service and well-being are four literature themes, with a glo-cal gap. It is important for the localised assessment.

4.1 Electrification in the MTF

As discussed earlier, the traditional definition of electrification is a binary measure only focusing on connections or non-connections, access and non-access and haves and haves not. It has been extended to reflect multi-dimensional issues under the MTF. Electrification is multi-dimensional when it captures the ability to avail energy that is adequate, available when needed, reliable, of good quality, affordable, legal, convenient, healthy and safe for all required services [2]. This shift in focus ensures that energy services of value to consumptive, productive and service related needs are targeted [2, 7]. Rwanda has already adapted its national statistics to reflect the MTF definition[4] [3]. For example, the electrical appliance use gap is highest in Tier 0 (zero distribution of appliances for 60.6 distribution of households across aggregated Tier- rural areas). Additionally, these households are in Ubudehe categories 1 and 2 and Social Classes 1 and 2 (**Tables 2–4**).

Though the MTF has been criticised for being complex to track at the global level and descriptive for acceptance at the national level [42], it has been deployed in other ways to improve the electrification experience[5]. One study constructed a bottom-up load profile at the household level for each tier of electricity access as set by the MTF, and the experiment was successfully tested in Rwanda, showing scalability [43]. The authors demonstrate further the inevitability of the present day Solar Home System (SHS) to meet energy demands beyond Tier 2 level and hence

	Tier 0	Tier 1	Tier 2	Tier 3	Tier 4	Tier 5
Distribution of households across aggregated Tier (Urban)	4.7	0.3	0.5	6.5	1.9	5.4
Distribution of households across aggregated Tier (Rural)	60.6	7.4	0.8	6.2	2.2	3.4

Source: Adapted from NISR [40].

Table 2.
Distribution of households across aggregated tier (urban/rural).

Appliance level	Tier 0	Tier 1	Tier 2	Tier 3	Tier 4	Tier 5
Type of appliances	—	Radio	Radio with CD Player and Mobile Phone	Electric Fan, TV, Computer and Printer	Refrigerator/ Freezer	Cooker
Distribution of appliances across Tiers	0	83.3	11.7	0.8	0.2	4

Source: Ref: [40].

Table 3.
Distribution of appliances.

[4] To this end, evidence from the Fifth Integrated Household Living Conditions Survey (EICV5) datasets show that the classic binary measure underestimates electrification rates by 8% (MTF, 35% and classic binary electrification rate, 27%) [40].

[5] While the MTF does not explicitly measure ownership of appliance, the implied ownership and usage of a set of typical appliances does play a significant role for dimensions such as peak power supply and activity levels used in establishing minimum daily energy supply threshholds [42].

Social class	Ubudehe category	Ubudehe explanation
Poor	Category 1	Families who do not own a house and can hardly afford basic needs
Low-income	Category 2	Those who have a dwelling of their own and or are able to rent one but rarely get full-time jobs
Medium-income	Category 3	Those who have a job and farmers who go beyond subsistence farming to produce a surplus that can be sold. The latter also includes those with small and medium enterprises who can provide employment to dozens of people
Wealthy	Category 4	Those who own large-scale businesses, individuals working with international organisations and industries as well as public servants

Source: Refs: [3, 7, 41].

Table 4.
Community categorisation (Ubudehe category and social class).

the need for sizing [43]. Using MTF interconnection of mini-grids and main-grids enhances reliability [9]. The energy poverty gap measured by supply and demand can also be explained through the MTF [9]. In the reference context, should supply be at the highest level of the tier, with a complementary low value measurement on consumption, there are two ways to explain the anomaly. One could be that there is an inability to pay and another is that there is low demand for high amounts. The latter case, concerns the unavailability of appliances or the availability of inefficient appliances. In this regard, a localised MTF assessment adopted in this study captures unavailable appliances and available inefficient appliances.

4.2 Decentralisation and centralisation

Electrification expansion in sub-Saharan Africa will happen through two main pathways; by expanding the existing grid and improving system efficiency to already existing customers; and by connecting new consumers through off-grid services. The binary measure of electrification from years 2000, 2005 and 2010 in comparison to year 2017 in sub-Saharan African countries in general shows improved electrification rates (see Appendix 1) [44]. However, rates continue to be low in rural areas compared to urban areas. Decentralisation of the grid through SHSs, mini-grids and other renewable energy based sources is receiving significant interest in remote and hard to reach areas which might never be reached through the grid or if the grid eventually reaches them it may be very costly and time-consuming. Around the 1930s, the United States of America (USA) was faced with a similar urban-rural electrification/appliance uptake gap. The Rural Electrification Administration (REA)'s strategies included travelling road shows (Electric Circus)-specifically designed to promote appliance uptake in rural farms[6]. The non-interest loan programme improved farm productivity in general [26] and an almost 100% electrification was achieved between 1930 and 1960 (a jump from 10%) and recorded both short-term and long-term growth [45]. To this end, resource constrained settings require targeted appliance uptake and energy demand stimulation packages suitable to both new and existing customers of which a localised MTF assessment adopted in this Chapter captures specific user needs.

[6] Between 1939 and 1941, representatives from the REA organised a carnivalesque roadshow designed to encourage families to purchase and use electrical appliances and other equipment in their homes and on their farms.

4.3 Energy as a service and the well-being context

Energy as a service is another important beyond electrification dimension focusing on well-being, for example, a recent study linked energy poverty, energy consumption, household level (patterns) and urban-rural development to the well-being context [15]. The authors describe this as the bottom-up perspective that includes the functions, services, benefits and values. Extending this, authors argue that benefits and values are regarded as qualitative indicators that depend on individual and cultural contexts. In the same context, mutual inference guides the relationship between values and benefits. For example, individual preferences and WTP are shaped by experiencing and by observation of the benefits derived from services. At the same time, existing values and moral stances influence how contributions to well-being are perceived. This is also the perceived value of appliances and electrification described in Uganda [24, 46–48]. Based on this observation, a localised MTF assessment described in this Chapter is crucial to capture well-being gaps and identification of functionalities and appliances of value to users.

4.4 Domestic end-use/electricity consumption

Domestic end-use electricity consumption has been characterised by technique (statistical/regression, engineering and NN methodologies, **Table 5**). According to McLoughlin et al. [49], statistical/regression models can be both bottom-up and top-down. They are bottom-up when data used is collected at an individual dwelling level. They are top-down when data is collected at an aggregate, for example, national energy statistics, and GDP and population figures. Statistical/regression models are useful when a large dataset exists as they are based on real data and give a good understanding of electricity consumption patterns. Engineering and NNs on the other hand, are bottom-up modelling approaches as they use data gathered at the dwelling level to infer relationships between electricity uses and dwelling and occupant characteristics. McLoughlin et al. [49], summarised that:

i. Statistical/Regression methods are costly to implement and sometimes suffer from multicolinearity between variables.

ii. Engineering models are the only methodology that can be used without any historical information on electrical use. However, they may be complex to implement and need to be validated.

iii. NNs can model complex input parameters and may provide accurate means of modelling, however, they can also suffer from multicolinearity.

In all the methodologies, appliance ownership is a key emergent variable of demand side characterisation. It is important that appliance uptake strategies receive adequate attention in literature and practice. A summary of key techniques from selected studies, see **Table 5** [27–39].

Further related works are summarised: Hamidi et al. [50], Diemuodeke et al. [51], Richardson et al. [52], Debnath et al. [53], Paatero and Lund [54], Cao et al. [55], Palensky and Dietrich [56], Firth et al. [57], Guerra-Santin et al. [58], Menezes et al. [59]. Hamidi et al. [50] proposed a generic approach to quantifying the level of responsiveness among domestic consumers, by deriving load appliances of target consumers. This approach benefits domestic consumers who have not yet benefited

Author	Methodology	Findings	Important metrics/data
STATISTICAL/ REGRESSION			
O'Doherty et al. (2008)	Papke-Wooldridge generalised linear model to infer a relationship between ***appliance ownership*** and electricity consumption	Explanatory variables that had a high significance for electricity consumption (dwelling characteristics, location, value and dwelling type as well as occupant characteristics; income, age, period of residency, social class and tenure type)	Used data from the Irish National Survey of Housing Quality (NSHQ) carried out in 2001–2002. The survey gathered information from a sample of over 40,000 householders on characteristics and problems of the dwelling, and on household members.
Parti and Parti (1980)	Conditional Demand Model (CDA)	A high significance of ***appliance ownership*** over electricity consumption patterns across a 24-hour period	Monthly electricity bills over a yearly period were regressed against appliance ownership figures and demographic variables such as household income and number of occupants to disaggregate electricity demand into 16 different end-uses
ENGINEERING			
Yao and Steemers (2005)	Dynamic software model to generate load profiles based on occupancy patterns, **appliance ownership** and ratings.	Categorised electricity consumption determinants based on two categories: behavioural and physical both of which are strongly related to dwelling occupancy patterns*.	A set of twelve monthly cross section regression analyses of the household demand for electricity was conducted
Widen and Wackelgard (2010)	Modelling framework for stochastic generation of time resolved data.	Authors found it an effective way to generate load profiles	Time-use data (.i.e. occupant's schedule of activities) as well as ***appliances*** holdings, ratings and day-distributions to produce electricity load profiles
Shimoda et al. (2004)	Simulation model (using all the households in Osaka city, Japan-divided into 460 types of dwellings)	Occupant's time-use, external temperature, **appliance efficiencies** and dwelling thermal characteristics significantly influenced the electricity consumption patterns per day	Modelled electricity consumption on an hourly basis for different dwelling and household characteristics
Capasso et al. (1994)	Modelled electricity consumption patterns at a 15 minute period,	Homeowner's occupancy patterns, as well as ***appliance ownership***, usage and ratings contributed to significantly constructing the load profile shapes	A model of electric residential use (Knowledge of its most relevant socioeconomic and demographic characteristics, unitary energy consumption and

Author	Methodology	Findings	Important metrics/data
			the load profiles of individual ***household appliances***; several probability functions, Monte Carlo extraction process and simulation)
NN			
Aydinalp et al. (2002)	Developed a NN	Modelling electricity consumption for ***domestic appliances***, lighting and space cooling in the home	NN methodology used in developing the appliances, lighting, and space-cooling component of the model, the accuracy of its predictions, and some sample results.
Aydinalp et al. (2004)	Extended Aydinalp et al., (2002) NN	Extended this work to develop NN models for space and domestic water heating	NN methodology extension
Aydinalp et al. (2008)	A comparison of NN conditional demand analysis and engineering approaches to modelling end-use energy consumption in the residential sector	Variables used in the NN model that influenced electricity consumption were ***appliance ownership*** and usage, income, dwelling type and household composition	NN methodology comparison

Behavioural determinants relate to decisions made on an hourly/daily/weekly basis regarding use of particular appliances. Physical determinants relate to fixed variables that do not change often or at all with time such as dwelling.
Source: Adapted from Ref: [49]; Additional Information from Systemic Review.

Table 5.
Approaches to modelling domestic electricity consumption.

from current systems. Diemuodeke et al. [51] employed a HOMER hybrid optimization software to determine the best solar energy system and recommended that it is efficient, cost effective, reliable, and environmentally friendly. Richardson et al. [52] used a domestic electricity demand model based on occupant time-use data, and noted that the model overlooked overnight demand; that people sometimes leave lights on while asleep or may use timers to run appliances. Cao et al. [55], used a two-stage budgeting framework and detailed micro-survey data to estimate energy demand system in urban China and found that poor households are sensitive to the price of coal and rich households are sensitive to the price of gasoline. Firth et al. [57] recorded five-minutely average whole house power consumption over 72 dwellings at five sites over 2 years and found an overall increase in electricity consumption attributable to a 10.2% increase in consumption of 'standby' appliances (televisions and consumer electronics) and a 4.7% increase in the consumption of 'active appliances' (lighting, kettles and electric showers). And that consumption of different energy user groups is low but high income users contribute to the overall increase in consumption.

A local level MTF assessment is crucial because users differ in their energy use behaviours and patterns. At the same time, appliances have different characterisations, depending on the study purpose or methodology in which they are being studied for. In Rwanda, Ugirimbabazi [12] applied the HOMER hybrid software to determine the best renewable based power system using a typical rural village load (the village load was adopted for further analysis and discussion from the appliance use perspective in this Chapter).

5. Conceptual framework

In this section, a characterisation of appliance uptake and energy consumption for transitioning economies is discussed with a specific contextualisation of the off-grid sector (low income and resource constrained settings). In transitioning studies, Wolfram et al. [19], found that:

- Economic growth will lead to large gains in residential sector energy use as households coming out of poverty purchase energy-using assets.
- Demand for electrical appliances will increase energy demand for rural dwellers who have yet to acquire even the most basic energy-using assets.
- Households coming out of poverty have much higher income elasticities of demand for energy-using assets.

The works of Bowden and Offer [25] and Wolfram et al. [19] use diffusion approaches to explain behavioural characteristics of appliance uptake. They inferred on the S-Curve to explain adoption of energy-using appliances through following their utility functions. Bowden and Offer [25] used the costs and benefits of discretionary time conceptual framework to explain why home entertainment (such as radio and TV) and kitchen machines (e.g. vacuum cleaners, washing machines and refrigerators) diffuse more quickly than others. They found diffusion of time-saving appliances as going ahead of income. As household income rise, consumers give time to their discretionary time. The authors defined time-saving goods as those reducing the time required to complete a specific task. Time-using goods are those which require the use of discretionary time, time which can be used according to the person's taste. Wolfram et al. [19] assessed cars and refrigerators diffusion across 28 countries in both the developed and the developing world by modelling the appliance or vehicle acquisition decision and adding features relevant to the developing world as follows:

> *The basic logic is straightforward. Households face a choice between consuming a divisible good with decreasing marginal utility (such as food) and an indivisible appliance that provides a fixed utility. As households' income increases, utility from increased consumption of the divisible good declines and, the probability that the household's utility from the appliance exceeds the utility from forgone food increases. Under reasonable assumptions on the distribution of appliance or vehicle valuations, this generates an S-shaped ownership curve* [19].

Appliance ownership is also low because most energy-using assets are expensive and most low-income households in the developing world are credit-constrained. A household does not make a period-by-period choice of whether to own an asset effectively by renting it, as is assumed in much of the developed-market literature. Instead the household must save to acquire the asset, which delays the asset acquisition to a higher income than would be suggested by the rental model.

Because lower income households are less able to self-finance, this delay is bigger at lower income levels and the resulting S-curve becomes steeper. Also, if households are self-financing through savings growth in income, and not just current income, will affect the asset acquisition [19].

From this conceptual framework, S-curve assumptions guide the Chapter as follows: In the first stages of adoption (poor households and medium-income households) will likely not have enough money to buy appliances. There are few initial purchases. This is the period that coincides with initial stages of development under the S-Curve. In the second stage of adoption, households have saved to afford purchases or can borrow. There is an exponential growth in purchases. Where there is access to roads and electricity, purchases of cooling appliances like refrigerator will increase. Additionally, households will purchase energy efficient appliances, including electric cook stoves. Wealthy households may buy expensive electric cook stoves while poor and medium income households may buy inefficient (cheaper) electric appliances. I assume that this is the stage that will coincide with rural Rwandan consumers beyond 2024 and the SEU model which underscores sustainable citizenry is introduced in Section 8 to explain this future.

6. Methods and data

6.1 Data

As discussed earlier, a macro-level dataset (EICV5) with 14,580 households has adapted the MTF definition (Section 4) [40]. I use this dataset to demonstrate tier characterisation at the national level. I complement the national level characterisation with village load data [12]. The village load has 164 users (**Table 6**). While the village load may not be a direct replica of all the villages in Rwanda, it provides a good starting point for the MTF localisation discussion from the appliance uptake perspective. Having compared appliance ownership and diffusion patterns in the off-grid market [7, 40], the village load adopted is a reasonable proxy, the source is credible academic work [12]. Use of an existing load profile is also time and cost effective. Difficulty in administering surveys (energy use and consumption based) and the associated uncertainty is documented [20, 60]. Data on newly created households is obtained from the Population and Housing Census of Rwanda (2012) (**Table 7**) [61].

6.2 Methods

6.2.1 Scenario analyses using payment mechanisms (a, b, c)

Scenario analyses explore WTP and ATP for any associated tarrif regime. These payment mechanisms are adopted from earlier studies elsewhere to demonstrate the role of different amounts of willingness (**Table 8**).[7]

[7] The selected payment mechanisms used are adopted from other contexts to demonstrate changes on the tarrif regime as different amounts are adopted.

User Classification by Ugirimbabazi (2015)	Category	Number	User classification adopted in this chapter
Domestic purposes	Rich families	10	Consumptive Sector
	Medium income families	40	Consumptive Sector
	Low income families	100	Consumptive Sector
Industrial/Commercial/ Community Purposes	Shops and bars	5	Productive Sector
	Administration posts	2	Service Sector
	Medical center	1	Service Sector
	Primary school	1	Service Sector
	Secondary school	1	Service Sector
	Community church	1	Service Sector
	Small manufacturing units	3	Productive Sector
Total		**164**	

Source: Ref: [12].

Table 6.
Composition of users for the village load.

Projections year	Rural Population	Mean size	Total households	Newly households to be created
2024	10,446,563	3.6	2,896,273	86,518
2025	10,581,467	3.5	2,985,468	89,195
2026	10,714,117	3.5	3,077,416	91,948
2027	10,844,122	3.4	3,172,174	94,758
2028	10,970,613	3.4	3,269,664	97,490
2029	11,092,996	3.3	3,369,885	100,221
2030	11,210,972	3.2	3,472,931	103,045
2031	11,324,247	3.2	3,578,902	105,971
2032	11,432,529	3.1	3,687,913	109,011

Source: Ref: [61].

Table 7.
Evolution of the number and size of the private households and the newly created private households between 2024 and 2032 by area of residence according to the medium projections scenario (rural).

6.2.2 Scenario analysis using energy consumption and newly created households (scenario d)

6.2.2.1 Basic model of energy consumption

As discussed earlier, domestic use predictions using linear models and theories assume exponential growth once households have saved enough money and can start buying appliances and demand energy use. This is also the case for diffusion theories discussed earlier. This Chapter uses a simplified method to capture energy

Scenario	Scenario a	Scenario b	Scenario c
What happens to the tarrif regime for different WTP and ATP (Scenario a, b and c)	WTP USD5.20/month	ATP USD 16.25/month	ATP USD 9/month
Data sources of the payment methodology (Scenario a, b and c)	*Ref:* [62]	*Ref:* [62]	*Ref:* [63]

Source: Ref: [62, 63].

Table 8.
Willingness and ability to buy.

demand (using a simplified relation between appliance, it's rating and duration of use) [64] as follows:

$$E_e = N_a \times A_r \times H_u \times P_n \quad (1)$$

Where:
E_e= energy use per appliance.
N_a= the number of appliances (of same kind).
A_r= the power rating of appliances in watts.
H_u= the duration of appliance usage (per day/365 days per year).
P_n= the number of households.

7. Results

7.1 Distribution of Rwandan households across Tier (National Level)

Tier 0 has the highest distribution of households across aggregated tier (**Table 9(a)**)
Distribution of households across capacity tier is also highest in Tier 0 (**Table 9(b)**)
Tier 5 has the highest distribution of households across duration tier-day time (**Table 9(c)**)
Distribution of households across duration tier-evening is highest in Tier 3 and Tier 5 (**Table 9(d)**)
Tier 5 has the highest distribution of households across reliability tier (**Table 9(e)**)
Distribution of households across quality tier is highest in Tier 5 (**Table 9(f)**)
All households meet the legality tier (**Table 9(g)**)
Tier 5 has the highest distribution of households across safety tier (**Table 9(h)**)

7.2 Village load

7.2.1 Appliance ownership

Appliance ownership is high for lighting appliances (lamps) and communication (cell phones and radio). Computers are common in the services sector (community church, secondary school, primary school, medical centre, administration post, shops and bars) (**Table 10**).

7.2.2 The village load and the multi-tier context

Based on the characterisation proposed in this study (consumptive, productive and services), the village load is distributable between Tier 2, 3 and 4. Thus there is

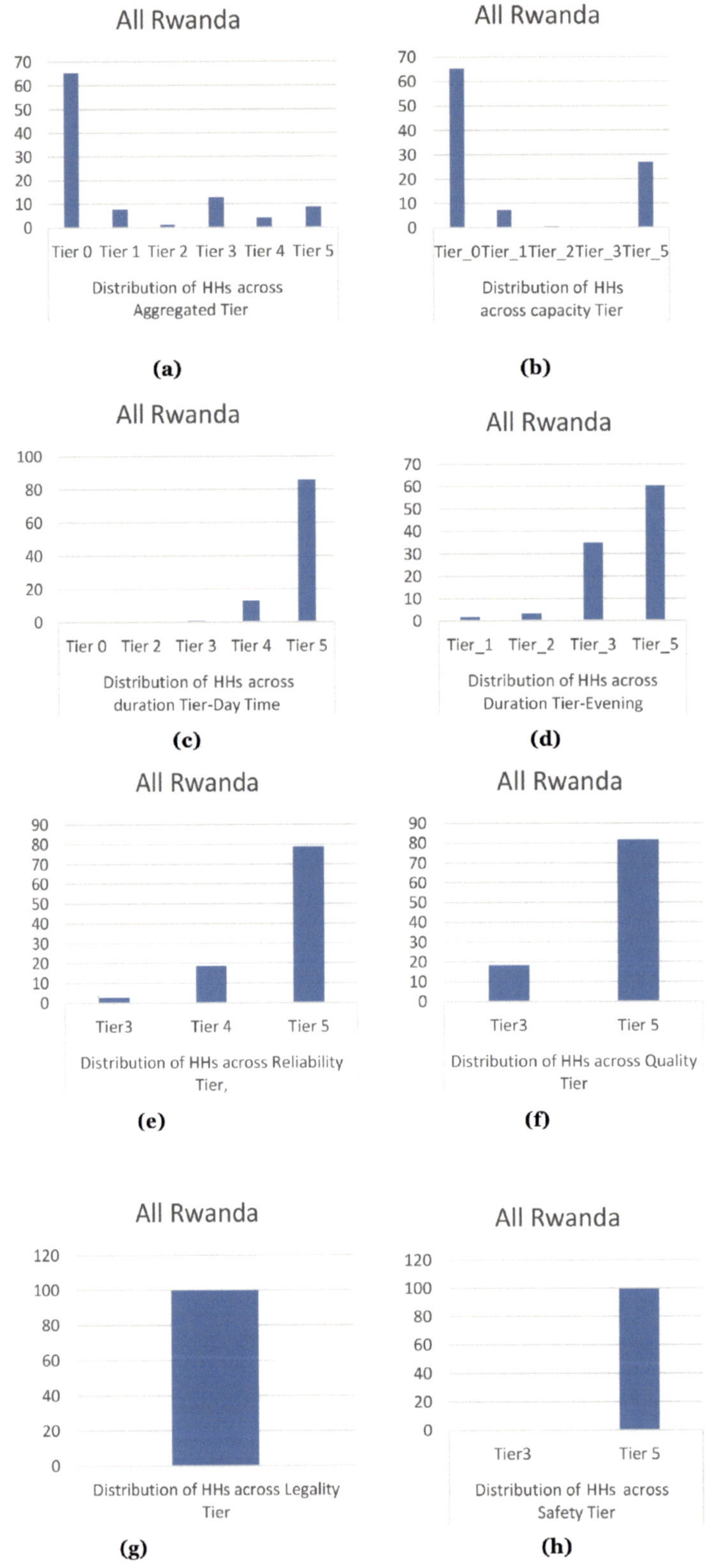

Table 9.
Distribution of Rwandan households across tiers (Source: Ref: [40]).

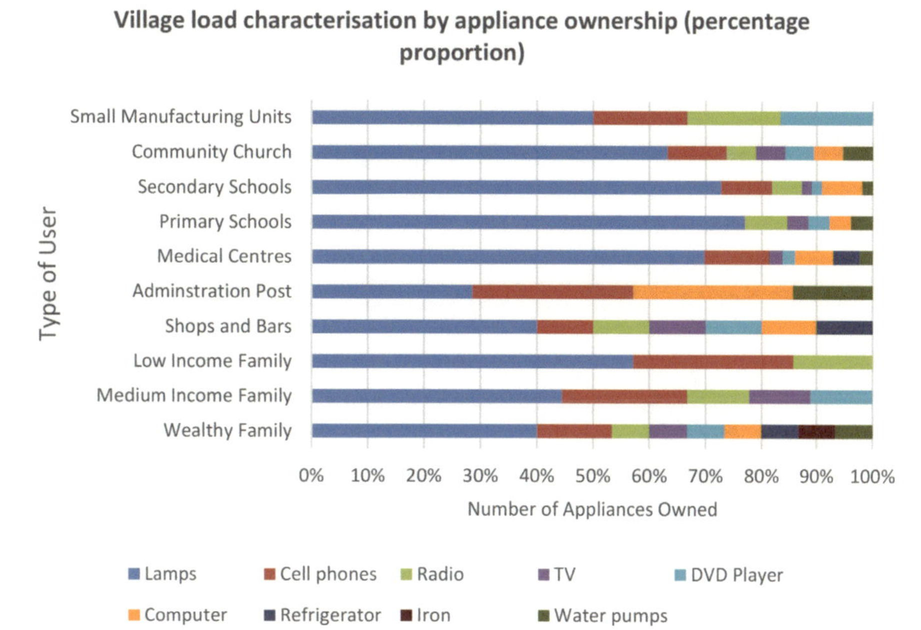

Table 10.
Characterising appliance ownership (percentage proportion) (Source: Ref: [12]).

	Tier 0	Tier 1	Tier 2	Tier 3	Tier 4	Tier 5
	Not applicable	Radio	Radio with CD Player and Mobile Phone	Tier 2 AND Electric Fan, TV, Computer and Printer	Tier 3 AND Refrigerator Freezer	TIER 4 AND Cooker
Wealthy Households					10 Families (46 kWh)	
Medium Income Households				40 Families (32kWh)		
Low Income households			100 Families (39kWh)			
Shops and Bars					5 Shops and Bars (35kWh)	
Administrative Post				2 Administration Posts (3kWh)		
Medical Center					1 Medical Centre (34kWh)	
Primary Schools				1 Primary School (5kWh)		
Secondary Schools				I Secondary School (11kWh)		
Community Church				I Community Church (5kWh)		

	Tier 0	Tier 1	Tier 2	Tier 3	Tier 4	Tier 5
Small manufacturing units			3 Small Manufacturing units (49kWh)			
Total			**88 kWh**	**56 kWh**	**115 kWh**	

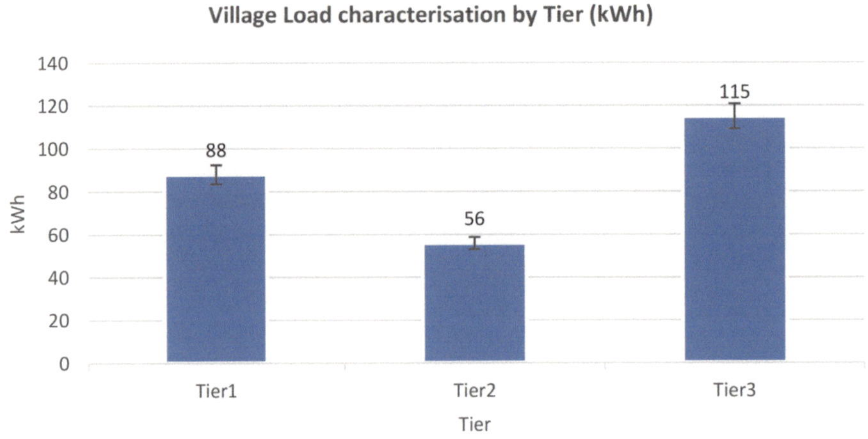

Table 11.
(a) Characterising village load; (b) characterising consumption by multi-tier context (kWh) (Source: Ref: [2, 12]).

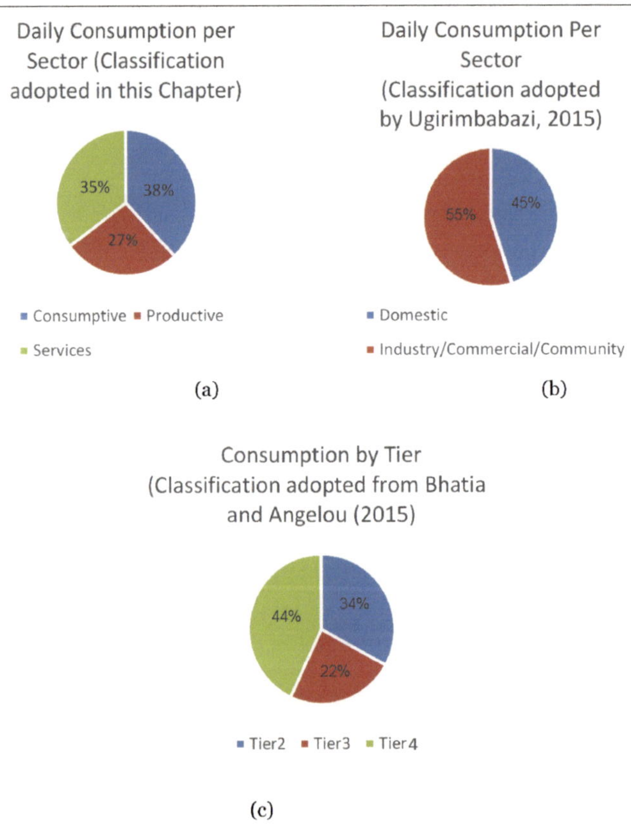

Table 12.
Characterising consumption by different classification (percentage) (Source: Ref: [2, 12]).

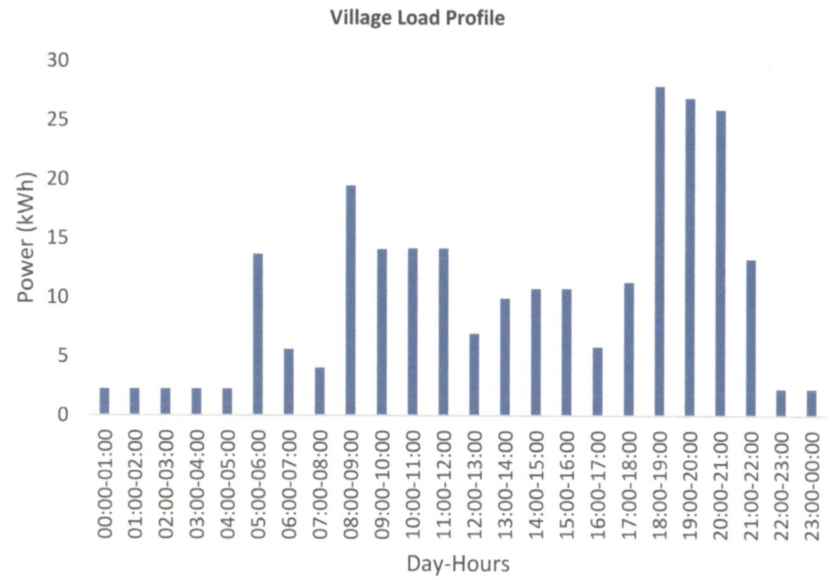

Table 13.
Hourly consumption (kWh) (Source: Ref: [12]).

no consumption for Tier 0 and Tier 5 (**Table 11**). Below we characterise total village consumption by different definitions such as (consumptive, productive and services adopted in this study 12(a)). The other characterisation is from the source document of the village load (12b) and the final one is the Tier approach. This section shows how definitional issues come to play in energy consumption, for example what is being counted and who is counting it. Also when was it counted [22] (**Table 12**).

7.2.3 Village load hourly consumption

The peak usage of energy is experienced between 19:00 and 22:00 hours and also between 09:00 and 12:00 hours (**Table 13**).

7.3 Scenario analyses (a-c)

Ceteris paribus, higher WTP and ATP by users yield higher tariffs. However, a high ATP is a business sustainability determinant than a high WTP (**Table 14**).

7.4 Scenario analysis (d)

7.4.1 Using daily consumption and newly created households

A further analysis of the household sector demonstrates that total energy use for the 2024–2032 period has an S-Curve pattern (Scenario d) (**Table 15**). It confirms Wolfram et al. findings of energy consumption behaviours of low income households coming out of poverty [19] and that as an economy grows its residential sector

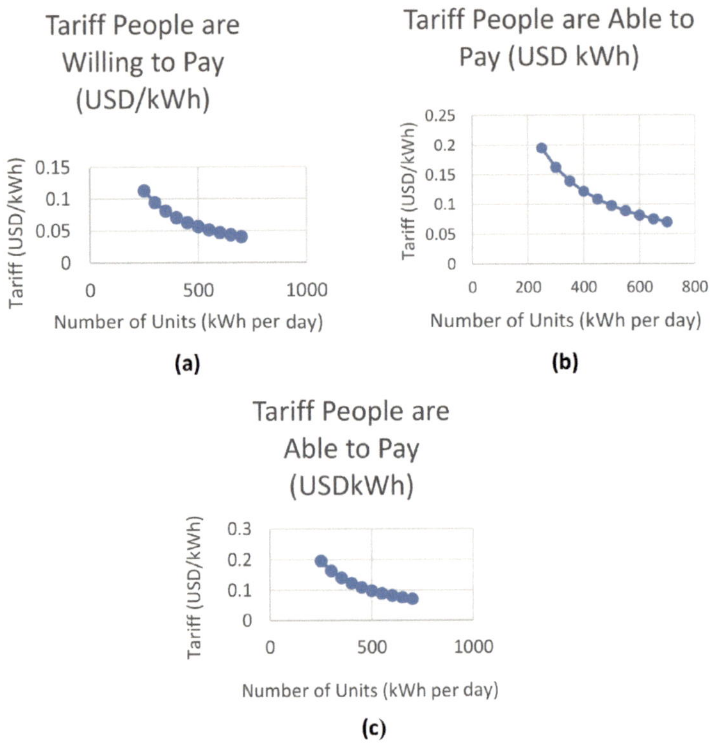

Table 14.
Tariff regimes and payment mechanisms (USD/kWh).

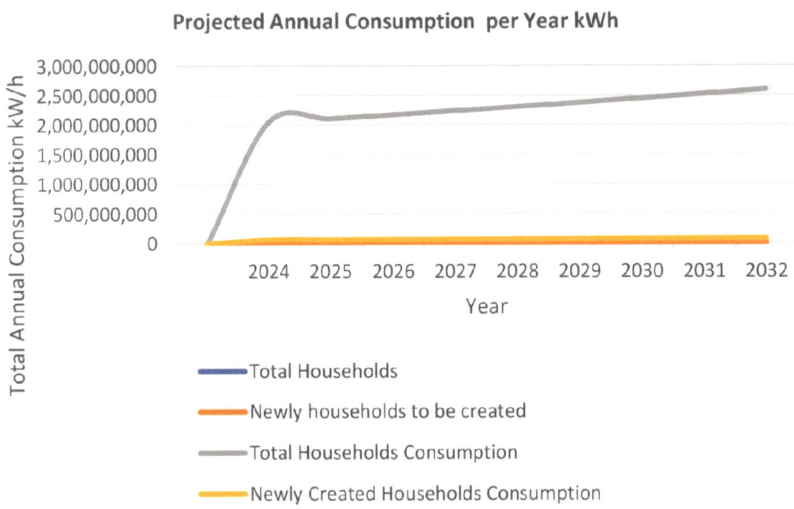

Table 15.
Scenario d (consumptive energy projection 2024–2032) (kWh/year) (Source: Author's Computations based on Ref: [12, 61]).

grows [21]. Energy consumption is heavily influenced by energy behaviours [66]. In Ghanaian urban households, high appliance ownership and usage is a key determinant of energy consumption [23].

8. Discussion and policy implications

8.1 Clean energy technologies and electrical appliances opportunities

As discussed earlier, insufficient energy consumption and appliance uptake deficiencies in the MTF context in this Chapter are discussed at the intersection of the theory of change and the theory of agency. In the case country, challenges include insufficient investments in energy infrastructures and the resultant energy crises (have negative effects on socio-economic development) [67]. Other constraints include (a) electricity demands almost equal with generation, with little reserves, (b) high petroleum products expenditures, (c) lack of investment, (d), government subsidies, which cushion electricity retail prices, and (e) inability to engage in much electricity export and trade because of relatively uncompetitive pricing regimes [68]. Rwanda's electricity price is about 22.2% more expensive than the highest tariff in the East African Region [68]. Moreover, high cost of electricity, generation capacity (demand and supply not aligned), insufficient resource margin and high system losses affect electrification prospects [69]. Feed-in-tariffs have been suggested until technologies are mature[8] [70].

Overally, external debt is reported as increasing electrification rates in the East African state [4]. Nonetheless, electrification rate reducers and increasers are distinguishable [4]. Rate increasers such as gross capital formation, external debt and agriculture. Rate reducers are multi-lateral debts and claims to central government. As per these conclusions by Mwizerwa and Bikorimana [4], I introduced the theory of change in this Chapter to underscore continued interventions by the Government of Rwanda (GoR) in reviving agriculture and promote women and youth inclusion (women and youth currently occupy 70% of the population). Improved outcomes of income and purchasing power may influence the capacity to buy electrical appliances. In another study conducted in Rwanda, connected households have more income compared to their matched unconnected counterparts [71]. To this end, electrification investments and feed-in-tariffs can cushion the negative transition effects and electricity utilisation strategies which should include complementary appliance use strategies to push forward off-grid electrification targets.

Related to the above discussion, the first pathway identified in this study, is participation of women and youth in productive use of energy technologies and appliances [26, 72]. As discussed earlier, agriculture as an electrification rate enabler (it is important that women and youth participation in productive use activities through the provision of energy and complementary energy-using appliances is promoted). Another study found appliance uptake to be highly gendered in Rwanda and the gender of the Head of Household (HoH) is a key driver in appliance uptake (from the Social Shaping of Technology perspective) [7]. At the same time, in terms of energy use, a study assessing productive use of electricity and street food in urban and peri-urban Rwanda found no significant difference between men and women (for all case study countries including Rwanda, Senegal and South Africa) [73]. However, Rwandan entrepreneurs had a preference for gas cookers and new appliances to attract customers to their businesses. In previous studies, specific differences in appliance uptake were found to be revolving around the use of discretionary time. For instance, time-using goods, are those which require the use of discretionary time in conjunction with the product [20]. Such trade-offs may be explained from the use of discretionary time perspective:

[8] This is mainly recommended for solar and wind technologies diffusion from the climate change adaptation and mitigation perspective.

- Radio and TV are typically time-using goods. They enhance perceived quality.
- Time-saving goods reduce the time required to complete a specific household task. While they are applied to housework, they can increase the quantity of discretionary time.

The second pathway concerns financing of appliances. This is because, a high WTP for electricity is not translating to the ATP cost recovering prices even under extended time periods [71]. A contigent behaviour analysis study in Rwanda determining potential benefits of electricity to unconnected customers shows that even the remotest customers are willing to pay for electricity [74]. However, the same study demonstrates that amounts customers are willing to pay cannot cover the cost of electricity which undermines the financial viability of projects [74]. Additionally, electrification benefits remain for minor use activities such as lighting, phone charging and agriculture processing. In the developing world, energy-using assets are expensive and low-income households are credit-constrained [19].

The third and final pathway considers economic activity stimulation. Three energy trends [55] from literature point to sectoral trade-offs that accompany energy transitions at the macro-level: as industrial energy demand increases most rapidly at the initial stages of development, growth slows steadily throughout the industrialisation process. Second, energy demand for transportation rises steadily, and takes the majority share of total energy use at the later stages of development. Third and finally, energy demand originating from the residential and commercial sector also increases to surpass industrial demand, but long term growth is not as pronounced in the transport sector. In this case, intersectoral linkages between infrastructure and industrial development will stimulate energy and appliances demand.

Other researchers found weak evidence of electrification on classical poverty indicators [21] and high energy bills [10] in Rwanda. In terms of energy efficiency, trade-offs exist between appliance uptake and associated benefits. For instance, anticipated benefits in improved lighting can be outweighed by uptake of other appliances like television. Differences in the speed of diffusion between appliances has been observed elsewhere, for example, going back in time, evidence from Britain and the USA, illustrates that since the 1920s, some household appliances diffuse more rapidly than others [20]. Home entertainment appliances such as radio and TV have diffused much faster than household and kitchen machines such as refrigerators. Differences in adoption are suggested to be a trade-off between energy-efficiency and cost of purchasing household appliances decision. To increase appliance uptake, ownership and use beyond electrification, further studies may explore this area. Recently, Sovacool [22] demonstrates that transitions appear not as an exponential line on a graph, but as a 'punctuated equilibrium which dips and rises'.

The appliance market across the globe has been observed as a niche. To this end, the 2020 global projections for the off-grid appliances in general, shows that fans, televisions and refrigerators are most promising with potential to reach $4.7B per year [8]. In Rwanda, this is relatable given the low village load across user groups or tiers (**Table 11**). Also, differing load profiles mean that as households transition from being low income to medium income to wealthy demand for appliances may improve. Evidence elsewhere shows high demand responsiveness in wealthy households, for example, in the UK wealthy households have at least one appliance under each use group; cold appliances- refrigerator, lighting appliances-light; brown appliances- TVs and Radio and Miscellaneous-Iron [50]. Both medium income and low-income households, only have brown appliances and lighting appliances [50]. Class distinction in appliance ownership is also observable in Rwanda (as discussed earlier). However, the minor distinction between low-income and medium-income households in terms of appliance ownership is an indicator of market potential. A consumer transition to wealthy status in Rwanda may stimulate appliance demand and energy use. In this transition it would be interesting to discuss the energy efficiency transition choices and preferences by consumers and implications for sustainable energy.

The S-Curve pattern demonstrated in Section 5 and 6 confirms consumer behaviours from world demand projections by earlier researchers (for example, [19]). As estimates by one study show that by 2035, developing world demand will almost double developed world demand [19]. Such an economic transition, will also mean, a consumer transition, as developing world customers become developed world customers. As households rise out of poverty and enter middle class category, they purchase new assets many of which use substantial amount of energy, and they also become, first time purchasers of energy-using assets [19].

Finally, deployment of Internet of Things (IOT) in energy studies has received significant interest elsewhere but also in Rwanda to address information asymmetries in the energy market. Particularly the use of large datasets in understanding consumer behaviour in energy markets. Kennedy et al. [75], used a BBOX database with 68600SHS customers over 562 days to compare non-parametric clustering method together with customer segmentation with linear models. Results demonstrate that linear models may be misleading because women and those recruited by agent advertising or word of mouth were more likely in the company's core clientele. Yet, linear models suggested that they are less profitable customers [75]. While IOT use is more likely to provide detailed insights on consumer behaviour to upscale business models, Bisaga et al. [76] notes that data privacy remains crucial. My follow on empirical work will determine emerging energy cultures using the Energy Cultures Framework and ground theory techniques. This will also be complemented by other forth-coming papers investigating willingness to use energy-using assets and the perceived user values (using the User Perceived Value-UPV Games and Questionnaire administration) and the resultant policy issues.

8.2 Why appliance efficiency?

In this section I discuss the reasons of appliance efficiency from two angles: first as a partial definition for sustainable energy and second as a reason for sustainable energy.

8.2.1 Appliance efficiency as a partial definition for sustainable energy

Labels and standards are regarded as valuable tools in implementing national energy efficiency policy [65]. Energy efficiency standards set minimum energy perfomance requirements for products and classes of products [65]. Labels are designed to inform consumer choice at the time of purchase and include endorsements, certifications, product comparisons, and product energy usage [65]. These valuable tools may assist in the avoidance of costly strategies such as China's Beijing refrigerator mistake (which contributed to the terrible air during the Beijing Olympics) [65].

Global savings in terms of energy efficiency are reported in terms of reduced energy use and costs 3–4% per year in all places where they have been introduced even in nations where they had no previous efficiency standards or program [65] (**Table 16**).

Metric	Labels/standards
80 Nations	Have adopted some kind of efficiency and/or labelling
55 different product types	Covered by a mandatory standard
3600 different policy measures	Addressing perfomance standards, and various forms of labelling
75 Nations	Have refrigerator measures
73 Nations	Regulate air conditioning
76 Countries	Have lighting measures
47 Countries	Have measures for television efficiency
Sources: Ref: [65].	

Table 16.
Global progress on appliance efficiency standards and labels.

8.2.2 Appliance efficiency as an example of sustainable energy

Specific benefits from appliance labels and standards are observable at four levels (individual, sectoral, national and international) (**Box 1**) [65]:

1. Individual: household level co-benefits (improved health and well-being, poverty alleviation, improved energy affordability and access and increased disposable income)

2. Sectoral: Industrial, transport, residential, commercial level co-benefits (include increased productivity and competitiveness, improved energy and other infrastructure benefits, and increased profits and asset values)

3. National: job creation, reduced energy-related public expenditure, energy security and valuable macro-economic benefits

4. International: Moderating energy prices, reducing natural resource pressure, and promoting the achievement of development goals

In developing countries, where light is provided either by candle or kerosene, or, if electricity is available, by an inexpensive incandescent bulb, new LED light technology could provide light using only 1 watt of power, which could be generated by a small solar panel and backed up by ordinary rechargeable batteries; total costs: US$25. These are examples of the astounding efficiency opportunities, that can be remotely powered without having to construct power plants or transmission lines, and that will provide light where none was before, or will eliminate GHG from burning fossil fuels or biomass. These LEDs are 1000 times more efficient at generating light than fuel based light (candles or kerosene), produce no indoor pollution, and have the potential, if they replaced fuel-based lighting, to save the equivalent of about 1.3 million barrels of oil per day. That would be a savings, at US$45 per barrel oil*, of about US$58.5 million per day—over US$ 21,352.5 billion per year—mostly in the poorest nations in the world, and virtually all of this money would be used to import the fuel. Reinvesting these savings in other energy efficient technologies could multiply the savings, while simultaneously improving the lives of over a billion people. And, as an added benefit, it would eliminate the 190 million tons of CO2 released annually when the fuel is burned.

**Price per barrel oil is adjusted to the 2019–2020 average of USD45.00 to provide the current situation.*

Source: Ref: [65].

Box 1.
One thought experiment demonstrating co-benefits of appliance efficiency.

8.3 Changing the energy consumption paradigm: SEU model

As discussed earlier a localised MTF assessment (Section 4) is crucial in addressing energy consumption related challenges of the Tier approach (**Table 1**). Empirical evidence demonstrated technologies and appliance uptake gaps (Section 3 and 4). Based on developed and developing world transition examples, energy transitions need to be complemented by matching business models. Recent studies have extensively explored the pros, cons and alternatives of different business models [16, 17]. This Chapter focused on a localised solution. The SEU model has positive attributes [77–81]. It stresses energy efficiency and renewable energy services to residents, businesses and governments. Demand for energy services jobs, sustainable energy services, financial savings that can accrue from efficiency investments (rather than relying on system benefit charges placed on utility bills for revenue) and a shared savings model for financing bonds to finance programs (**Figure 1**).

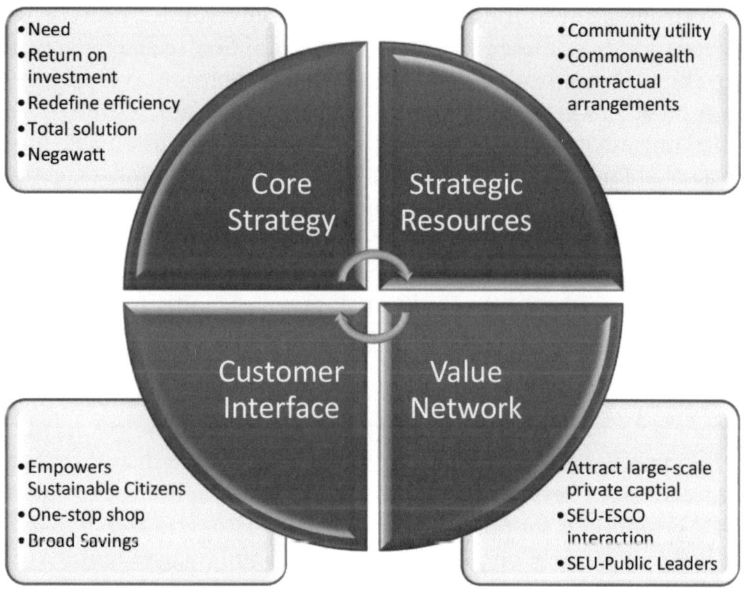

Figure 1.
The SEU model. Source: Ref: [17].

8.4 Policy implications and future work

This chapter demonstrated relevance of the utility of the bottom-up "polycentric" approach to off-grid rural electrification and its implications on the off-grid energy market in the developing world. Two theories: agency and change were used to assess the uptake of off-grid appliances in Rwanda, and the attendant direct relationship to clean energy investments and off-grid policy development. This is crucial given the relevance of both the MTF and energy transitions in today's sustainability discourse and the broader climate change objectives. While end-user awareness and participation in policy and business model development are necessary for increased rural electrification, community-based energy planning may have additional positive effects. Several key questions pertaining to off-grid clean energy and electrical appliances uptake, ownership, use and value include; When to increase investments? When to strengthen markets? How to do this? What is more preferable to consumers and Why?

Literature evidence and the scenario analyses in the current study suggests that energy consumers purchase decisions are usually informed by utility metrics such as customers' WTP and the ATP of which to strike a purchase, the ATP should always be higher than the WTP [82]. However, beyond ATP and WTP metrics, there are some exogenous factors that will influence the customer's decision to buy an appliance. The adoption decision depends on the product, individual and the environment [83]. In India, [5], improving the quality of service was identified as crucial for increasing the electricity price (for example, hours of supply per day). Gendered ranking of appliances use needs better informs energy planning [18]. Li et al. [72], recommends a consideration of multiple factors, including local energy resources and economic, social, cultural and national geographical factors influencing, the rural household energy consumption structure [72]. Appliances may also have different values for instance: functional, social significance, epistemic, emotional and cultural [46, 47].

Three major themes have characterised energy governance and policy planning over the last decade: fragmentation, complexity and polycentricity, and to enhance effectiveness of off-grid appliance policies, government decision-makers and firms need to address these issues in the context of new technology, markets, and policy innovations at multiscale levels. The utility of a bottom-up ("polycentric) innovation approach to the off-grid appliance space and related development policy practices/planning are imperatives [16, 17]. Such innovative sustainable business models are availed for subsequent diffusion across different countries, contexts and domains. They demonstrate the value of polycentric climate governance in the investigation of the sustainable business model innovation [17]. Evidence from this Chapter shows that integrating appliance use, preferences and values in bottom-up consumer perspectives in appliance policy planning is one way to de-risk markets.

Supply decisions will thus be informed by Sustainable User preferences, user trends, and the user value attached to appliances. For example, adoption rates of SHS in Central East Africa, followed 3 phases (phase A, B and C) [84]. In phase A, distributors lacked a marketing strategy and most sales were garnered through the shops. In phase B, sales experienced an exponential growth as distributors adopted aggressive marketing strategies such as recruiting local sales agents and running local promotions for both urban and rural customers and in phase C, the distributor had halted down on promotional events and focuses more on customer services.

8.4.1 Derisking investments, climate change, investment and policy innovations

8.4.1.1 Investments

The major gap in existing financial incentives and strategies for elimination of market distortions strategies is centralised planning and a supplier-focus. East African countries adopted different strategies. Kenya strengthened its on-grid investment complemented by early support for off-grid. Whereas, Tanzania has low regulation on niche innovators enabling off-grid projects [85]. Fee-for-services, financial incentive and collaborative local efforts (local arrangements) are possible bottom up financial mechanisms. It would be important to explore their viability in promoting clean energy technologies and electrical appliances uptake, ownership and use in Rwanda. Options include payment strategies such as cost-sharing, hire purchase, renting or appliance financing which can be further explored from the user's perspective.

8.4.1.2 Climate change adaptation and mitigation

Previously, capacity building and support to negotiators and local institutions were recommended from the negotiators (including training and logistical support equipment [65, 86]. At the global policy level, gaps noted were: (i) intra-generational/equity where few Clean Development Mechanism (CDM), projects were implemented in Africa, (ii) design flaw in the United Nations Framework Convenction on Climate (UNFCC), because they failed to allocate emission rights to all countries (iii) a proper enabling environment where market based mechanisms could ensure funds flow from big-emitters to low emitters [86]. Finally, policy interventions to eliminate factors that constrain the operation of climate change mitigation related to private sector investment in poor countries. At the local level, appliance efficiency as a cheap pathway to low carbon emissions and its integration to local development initiatives is crucial. For example, decarbonisation strategies of the second meeting of the Conference of the Parties to the Minamata Convention on Mercury (COP2) are already 'ground-up'. Feed-in-tariffs were discussed earlier until technologies are mature and also user engagement (particularly key stakeholders using high carbon technologies).

8.4.1.3 Government policy and derisking country-risk

While the off-grid market has been left to private investors in most African countries, overall management of the energy policy including renewables is centrally managed by public utilities of which that is not a bad thing. Empirical evidence shows that there are governments that have preference for government control in the overall management of energy planning and great strides in electrification for instance Ethiopia and Tanzania. However, it is not clear how the appliance uptake strategy is managed alongside electrification rates. For some governments, private enterprises are extensively encouraged to participate in electrification, for instance Kenya, Nigeria and Ghana. Again the clean energy technologies and electrical appliances uptake strategy is not well articulated.

In Rwanda, the energy utility manages the energy plan, but the role of the private sector is explicitly stated and the launch of mini-grids and SHSs standards is a great step. However, an appliance uptake strategy is imperative and so are the specific initiatives and support mechanisms. Given the massive energy demand and appliance uptake projected in the off-grid markets there is need for GOR to further articulate consumptive, productive and service oriented support mechanisms of the energy and appliance uptake transition. Next steps on the policy framework could include measuring functional perfomance of technologies to adequately address arising issues on appliance loads [87].

8.4.1.4 Private enterprises nudges

Minigrids capacity to promote economic empowerment activities in households, small scale enterprises and other high consumption activities is a limitation in old designs. As private players transition from old designs to new designs, feed-in-tariffs may be suggested. Additionally, the 'arrival of the grid' is feared to disrupt off-grid businesses. In areas where the grid will eventually reach, there is need for clear indication on how they will be connected to the grid and that private investors will still recover their investments. Rwanda has clearly stated its full support to off-grid market players. Specific interventions highlighted in the Rural Electrification Strategy [41] include technical assessments and siting incentives by its private sector players also marked on an off-grid map. However, participation of local stakeholders in technical assessments and siting activities and the general

Pathways	Demand Stimulation Packages
Gendered pathways	Appliances for value-added agriculture Specification of the role of women in electrical appliance uptake strategies Considering the role of the HoH in appliance use Adoption of technologies and appliances for agriculture productivity Designing an appliance uptake strategy for the modern Rwandan household and role of women in energy efficiency Identification of women's time use and time saving appliances Skills and capacity building across Tier requirements
Financing pathways	Cost-sharing in appliance purchase Investments for appliance financing Micro-credit/Loans Private sector players finance schemes
Economic Activity and Energy Use Stimulation Pathways	Increasing production on existing agriculture land Energy efficient appliances Strengthening the consumptive-productive-service sector linkages Exploring appliance initiatives in the Made in Rwanda campaign and small scale enterprises (Agakiriro activities) Organise farmers-credit cooperatives/agriculture cooperatives Set up a division to promote electricity demand in cooperatives Sharing information about electricity uses Advocating for quality certified products Working with appliances companies to target the rural market Public equipment demonstrations Appliance campaigns

Source: Author.

Table 17.
The electric circus and its pathways in rural Rwanda.

development plans for local consumers' may be fully considered. Though a specific off-grid map shows potential sites, consideration of consumer needs and desires in new sites may strengthen business models.

8.4.1.5 The potential of the electric circus

In summary the potential of electrification and electric appliances as was the case of the USA in the 1930s is a great case study for present day transitioning economies like Rwanda. The three pathways of a potential demand stimulation of energy and electrical appliances (**Table 17**).

Finally, the Chapter used a demand side characterisation approach to highlight systemic opportunities for stimulating appliance and energy demand at the intersection of the theory of change and the theory of agency. Three pathways emerging from the study focused on three themes: improving women and youth participation in productive use of energy and appliances, appliance financing and economic activity stimulation. However, as economies develop and experience economic growth a shift in the energy consumption approach is pertinent. The SEU model which has a transformative agenda underscores a consumer transition that leads to sustainable citizens. Its business model was used to influence derisking decisions which can influence investments, markets and policy innovations in a futuristic Rwanda. The pathways for a Rwandan Electric Circus were outlined.

Acknowledgements

I am grateful to the Editor and all the Editorial team and the assistance by many fine scholars too many to mention in this short work.

Appendix 1: Electricity Access in Africa, 2000–2017

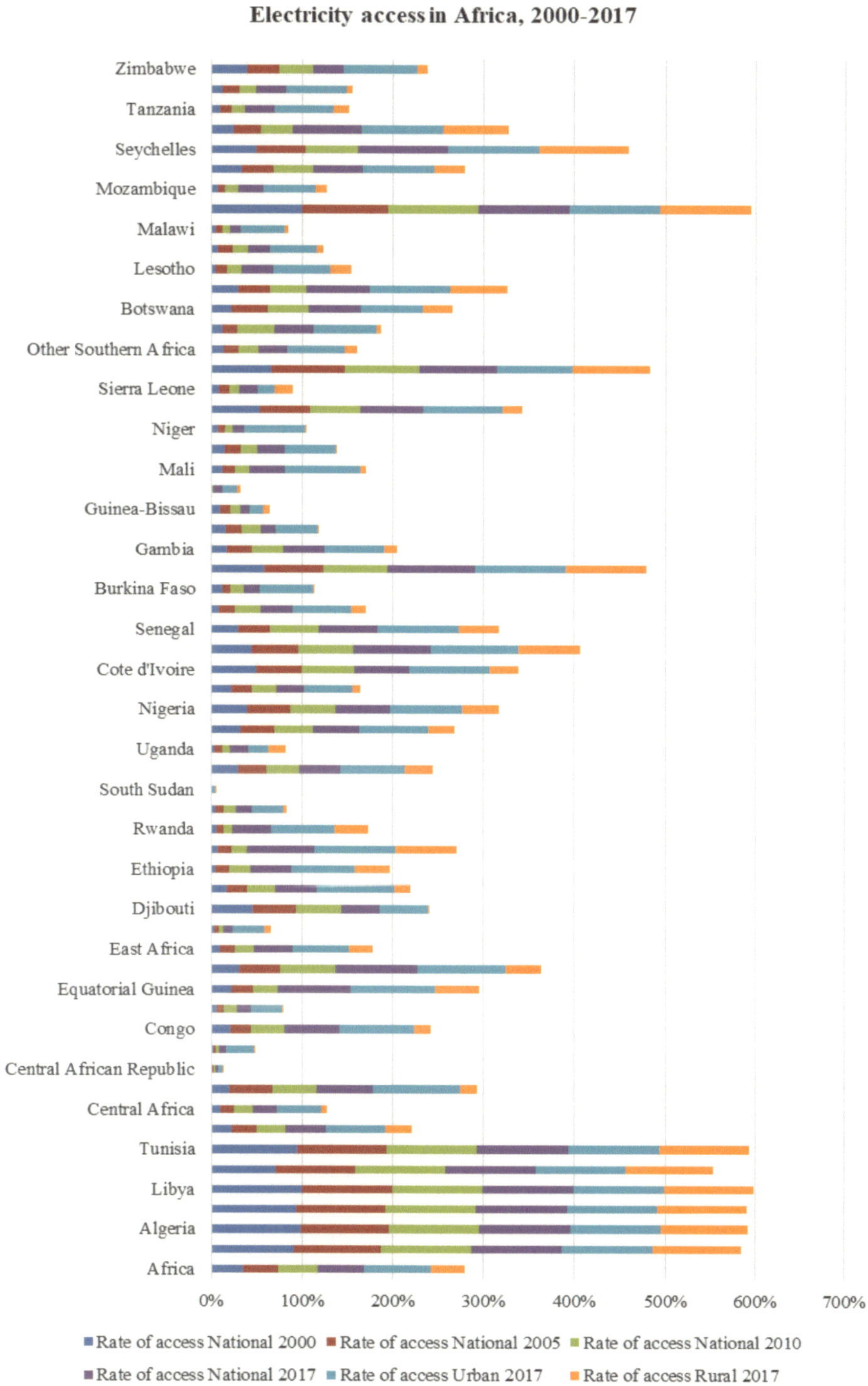

Source: Ref. [49].

Author details

Olivia Muza
African Centre of Excellence in Energy for Sustainable Development (ACE-ESD), College of Sciences and Technology (CST), University of Rwanda, Kigali, 4285, Rwanda

*Address all correspondence to: oliviamuza@gmail.com

IntechOpen

© 2020 The Author(s). Licensee IntechOpen. Distributed under the terms of the Creative Commons Attribution - NonCommercial 4.0 License (https://creativecommons.org/licenses/by-nc/4.0/), which permits use, distribution and reproduction for non-commercial purposes, provided the original is properly cited.

References

[1] L. Lenz, A. Munyehirwe, J. Peters and M. Sievert, "Does large-scale infrastructure investment alleviate poverty? Impacts of Rwanda's electricity access roll-out program," *World Development,* no. 89, pp. 88–110., 2017.

[2] M. Bhatia and N. Angelou, "Beyond Connections: Energy Access Redefined.," © World Bank., Washington, D.C, 2015.

[3] Republic of Rwanda, "Energy Sector Strategic Plan 2018/19—2023/24. September 2018. Ministry of Infrastructure. https://www.mininfra.gov.rw/fileadmin/user_upload/infos/Final_ESSP.pdf. Last accessed December 2019".

[4] D. Mwizerwa and G. Bikorimana, "Macroeconomic Determinants of Electricity Access in Rwanda, an Empirical Analysis," *Microeconomics and Macroeconomics,* vol. 6, no. 1, pp. 20–31, 2018.

[5] R. Kennedy, A. Mahajan and J. Urpelainen, "Quality of service predicts willingness to pay for household electricity connections in rural India," *Energy Policy,* vol. 129, pp. 319–326., 2019.

[6] M. Bergen, S. Dutta and O. J. J. Walker, "Agency Relationships in Marketing: A Review of the Implications and Applications of Agency Related Theories," *Journal of Marketing,* vol. Volume 58 (July 1992), pp. 1–24, 1992.

[7] O. Muza and R. Debnath, "Socially inclusive renewable energy transition in sub-Saharan Africa: A social shaping of technology analysis of appliance uptake in Rwanda," *Cambridge Working Paper in Economics 2055,* 2020.

[8] Global Leap, "The State of THE Off-Grid Appliance Market. Lighting and Energy Access Partnership.," Lighting and Energy Access Partnership., 2018.

[9] S. Groh, S. Pachauri and N. D. Rao, "What are we measuring? An empirical analysis of household electricity access metrics in rural Bangladesh," *Energy for Sustainable Development,* no. 30, pp. 21–31, February 2016.

[10] G. Bensch, "Rural Electrification in Rwanda – An Impact Assessment Using Matching Techniques," *ISBN 978-3-86788-264-4, Rheinisch-Westfälisches Institut für Wirtschaftsforschung (RWI),* 2010.

[11] G. Bensch, J. Kluve, J and J. Peters, "Impacts of rural electrification in Rwanda. Journal of Development Effectiveness, 3(4), 567–588.," *Journal of Development Effectiveness,* vol. 3, no. 4, pp. 567–588., 2011.

[12] O. Ugirimbabazi, "Analysis of Power System Options for Rural Electrification in Rwanda.(Master's thesis, Universitetet i Agder; University of Agder).," 2015.

[13] J. Uwibambe, "Design of Photovoltaic System for Rural Electrification in Rwanda (Master's thesis, Universitetet i Agder; University of Agder).," 2017.

[14] N. Mohlakoana, J. de Groot, A. Knox and J. T. Bressers, "Productive Uses of Energy: Gender and the Informal Sector.136–137. Paper presented at The 6th World Sustainability Forum, Cape Town, South Africa.," in *Paper presented at The 6th World Sustainability Forum, Cape Town, South Africa.*, 2017.

[15] G. Kalt, D. Wiedenhofer, C. Görg and H. Haberl, "Conceptualizing energy services: A review of energy and well-being along the Energy Service Cascade," *Energy Research & Social Sciences,* no. 53, pp. 47–58, 2019.

[16] J. Nyangon, "International Environmental Governance: Lessons

from UNEA and Perspectives on the Post-2015 Era," *Journal on Sustainable Development Law and Policy,* vol. 4, pp. 174–202, 2014.

[17] J. Nyangon and J. Byrne, Electricity Customer Choice: REVing Up the New York Energy Vision for Polycentric Innovation. In Energy System and Environment, Tsvetkov, P., Ed., London: IntechOpen, 2018.

[18] B. Koo, D. Rysankova, E. Portale, N. Angelou, S. O. Keller and G. Padam, "Rwanda-Beyond connections: energy access diagnostic report based on the multi-tier framework (English)," World Bank Group, Washington D.C., 2018.

[19] C. Wolfram, O. Shelef and P. Gertler, "How will energy demand develop in the developing world?. Journal of Economic Perspectives," *Journal of Economic Perspectives,* vol. 26, no. 1, pp. 119–38, 2012.

[20] S. Bowden and A. Offer, "Household appliances and the use of time: the United States and Britain since the 1920s.," *Economic History Review,* no. 4, pp. 725–748., November 1994.

[21] B. Medlock and R. Soligo, "Economic Development and End-Use Energy Demand," *The Energy Journal,* vol. 22, no. 2, April 2001.

[22] B. K. Sovacool, "How long will it take? Conceptualising the temporal dynamics of energy transitions. http://dx.doi.org/10.1016/j.erss.2015.12.020," *Energy Research and Social Science,* vol. 13, pp. 202–215, 2016.

[23] M. Sakah, S. D. L. R. du Can, F. A. Diawuo, M. D. Sedzro and C. Kuhn, "A study of appliance ownership and electricity consumption determinants in urban Ghanaian households," *Sustainable Cities and Society,* vol. 44, pp. 559–581, October 2018.

[24] S. Hirmer and P. Guthrie, "The benefits of energy appliances in the off-grid energy sector based on seven off-grid initiatives in rural Uganda," *Renewable and Sustainable Energy Reviews,* vol. 79, pp. 924–934, November 2017.

[25] S. Bowden and A. Offer, "Household Appliances and the Use of Time: The United States and Britain Since the 1920s," *JSTOR,* vol. 47, p. 24, November 1994.

[26] C. Kitchens and P. Fishback, "Flip the Switch: The Impact of the Rural Electrification Administration 1935–1940," *The Journal of Economic History,* vol. 75, no. 4, pp. 1161–1195, December 2015.

[27] J. O'Doherty, S. Lyons and R. S. Tol, "Enegy-Using appliances and energy-saving features:determinants of ownership in Ireland,Applied Energy," *Applied energy,* vol. 85, no. 7, pp. 650–662, 2008.

[28] E. Leahy and S. Lyons, "Energy use and appliance ownership in Ireland," *Energy Policy,* vol. 38, no. 8, pp. 4265–4279, Febraruary 2009.

[29] M. Parti and C. Parti, "The total appliance specific conditional demand for electricity in the household sector," *Bell Journal of Economics,* vol. 11(1), pp. 309–324, 1980.

[30] Y. G. Yohanis, J. D. Mondol, A. Wright and B. Norton, "Real-life energy use in the UK: How occupancy and dwelling characteristics affect domestic electricity use. Energy and Buildings, 40 (6), 1053–1059," *Energy and Buildings,* vol. 40, no. 6, pp. 1053–1059, 2008.

[31] M. Hart and R. De Dear, "Weather sensitivity in household appliance energy end-use," *Energy and Buildings,* vol. 36, no. 2, pp. 161–174, August 2004.

[32] D. S. Parker, "Research highlights from a large scale residential monitoring study in a hot climate, energy and

buildings," *Energy and buildings,* vol. 35, no. 9, pp. 863–876, 2003.

[33] R. Yao and K. Steemers, "A method of formulating energy load profile for domestic buildings in the UK," *Energy and Buildings,* vol. 37, no. 6, pp. 663–671, January 2005.

[34] J. Widén and E. Wäckelgård, "A high-resolution stochastic model of domestic activity patterns and electricity demand," *Applied Energy,* vol. 86, no. 6, pp. 1880–1892., June 2010.

[35] Y. Shimoda, T. Fujii, T. Morikawa and M. Mizuno, "Shimoda, Y., Fujii, T., Morikawa, T., & Mizuno, M. (2004). Residential end-use energy simulation at city scale," *Building and environment,* vol. 39, no. 8, pp. 959–967, 2004.

[36] A. Capasso, W. Grattieri, R. Lamedica and A. Prudenzi, "A bottom-up approach to residential load modeling," *IEEE Transactions on Power Systems,* vol. 9, no. 2, pp. 957–964, April 1994.

[37] M. Aydinalp, V. I. Ugursal and A. S. Fung, "Modeling of the appliance, lighting, and space-cooling energy consumptions in the residential sector using neural networks," *Applied Energy,* vol. 71, no. 2, pp. 87–110, 2002.

[38] M. Aydinalp, V. I. Ugursal and A. S. Fung, A. S, "Modeling of the space and domestic hot-water heating energy-consumption in the residential sector using neural networks," *Applied Energy,* vol. 7, no. 2, pp. 159–178,, October 2004.

[39] M. Aydinalp-Koksal and V. I. Ugursal, "Comparison of neural network, conditional demand analysis, and engineering approaches for modeling end-use energy consumption in the residential sector," *Applied energy,* vol. 85, no. 4, pp. 271–296, April 2008.

[40] Republic of Rwanda, "EICV5. Available on: https://www.statistics.gov.rw/datasource/integrated-household-living-conditions-survey-5-eicv-5, Date Accessed 11 April 2019," NISR, 2016/2017. [Online].

[41] Republic of Rwanda., "Rural Electrification Strategy. Available on: https://www.mininfra.gov.rw/fileadmin/user_upload/aircraft/Rural_Electrification_Strategy.pdf. Last accessed on December 2019. Ministry of Infrastrucutre," 2016. [Online]. [Accessed May 2016].

[42] S. Pelz, S. Pachauri and S. Groh, "A critical review of modern approaches for multidimensional energy poverty measurement.," *Energy and Environment,* vol. 7, no. 6, e304, October 2018.

[43] N. Nayaran, Z. Qin, J. Popovic-Gerber, J.-C. Diehl, P. Bauer and M. Zeman, "Stochastic load profile construction for the multi-tier framework for household electricity access using off-grid DC appliances," *Energy Efficiency,* vol. 240, no. 12:185, pp. 907–917, August 2019.

[44] IEA, "World Energy Outlook. Available on https://www.iea.org/reports/world-energy-outlook-2018. Last accessed on 25 September 2019," 2018.

[45] J. Lewis and E. Severnini, "Short- and long-run impacts of rural electrification: Evidence from the historical rollout of the US power grid," *Journal of Development Economics,* vol. 143, no. 102412, 2020.

[46] S. Hirmer and H. Cruickshank, "The user-value of rural electrification: An analysis and adoption of existing models and theories," *Renewable Energy and Sustainable Energy Reviews,* vol. 34, pp. 145–154, June 2014.

[47] S. Hirmer and P. Guthrie, "Identifying the needs of communities in rural Uganda: A method for determining the 'User-Perceived

Value'of rural electrificationinitiatives," *Renewable and Sustainable Energy Reviews,* vol. 66, pp. 476–486, October 2016.

[48] Hirmer and S, Improving the Sustainability of Rural Electrification Schemes: Capturing Value for Rural Communities in Uganda. (Ph.D. Thesis, University of Cambridge), Cambridge: University of Cambridge, 2018.

[49] F. McLoughlin, A. Duffy and M. Conlon, "Characterising domestic electricity consumption patterns by dwelling and occupant socio-economic variables:An Irish case study," *Energy and Buildings,* no. 48, pp. 240–248, January 2012.

[50] V. Hamidi, F. Li and F. Robinson, "Demand response in UK's domestic sector," *Electric Power Systems Research,* vol. 79, pp. 1722–1726, August 2009.

[51] E. O. Diemuodeke, A. Addo, I. Dabipi-Kalio, C. O. Oko and Y. Mulugetta, "Domestic energy demand assessment of coastline rural communities with solar electrification," *Energy and Policy Research,* vol. 4, no. 1, pp. 1–9, 2017.

[52] I. Richardson, M. Thomson, D. Infield and C. Clifford, "Domestic electricity use: A high-resolution energy demand model," *Energy and Buildings,* vol. 42, no. 10, pp. 1878–1887, July 2010.

[53] K. B. Debnath, M. Mourshed and S. P. K. Chow, "Modelling and forecasting energy demand in rural households of Bangladesh. Energy Procedia, 75, 2731–2737.," *Energy Procedia,* vol. 75, pp. 2731–2737., 2015.

[54] J. V. Paatero and P. D. Lund, "A model for generating household electricity load profiles," *International Journal of Energy Research,* vol. 30, no. 5, pp. 273–290, April 2006.

[55] J. Cao, M. S. Ho and H. Liang, "Household Energy Demand in Urban China: Accounting for Regional Prices and Rapid Income Change," *The Energy Journal,* vol. 37, no. China Special Issue, pp. 87–110, 2016.

[56] P. Palensky and D. Dietrich, "Demand Side Management: Demand Response, Intelligent Energy Systems, and Smart Loads," *IEEE Transactions on Industrial Informatics,* vol. 7, no. 3, September 2011.

[57] S. K. Firth, K. J. Lomas, A. J. Wright and R. Wall, "Identifying trends in the use of domestic appliances from household electricity consumption measurements," *Energy and Buildings,* vol. 40 (5), pp. 926–936, 2008.

[58] O. Guerra-Santin, H. Bosch, P. Budde, T. Konstantinou, T. Boess, T. Klein and S. J. E. E. Silvester, "Considering user profiles and occupants' behaviour on a zero energy renovation strategy for multi-family housing in the Netherlands," *Energy Efficiency,* vol. 28, pp. 311–327, June 2018.

[59] A. C. Menezes, A. Cripps, R. A. Buswell, J. Wright and D. Bouchlaghem, "Estimating the energy consumption and power demand of small power equipment in office buildings," *Energy and Buildings,* vol. 75, pp. 199–209, February 2014.

[60] T. Inaba, T. Tabata and T. Peii, "Development of a Basic Rate of Household Energy Consumption Considering Usage Time and Quantity of Consumer Durables," *Journal of Sustainable Development of Energy, Water and Environment Systems,* vol. 5 (4), pp. 533–545, 2017.

[61] Republic of Rwanda, "Population and Housing Census of Rwanda, 2012. Last accessed on 01 March 2019 on: h ttps://www.statistics.gov.rw/data source/population-and-housing-census. NISR 2012".

[62] IFC, "Mini-grid market assessment Kenya 2015. Available on:https://www.usaid.gov/sites/default/files/documents/1860/PAOP-Market-Assessment-Brief-Kenya-English.pdf. Last Accessed on 10 January 2020," 2015.

[63] DFID-IED, "Low Carbon Mini-Grids. Identifying the Gaps; Building the Evidence: Volume 1. Available on:http://documents1.worldbank.org/curated/en/792001512392701402/pdf/ESM-cKenyaMiniGridsCaseStudyConfEd-PUBLIC.pdf, Last Accessed on 10 January 2020," 2013.

[64] E. A. Essah and E. L. Ofetotse, "Energy supply, consumption and access dynamics in Botswana," *Sustainable Cities and Society,* vol. 12, pp. 76–84, 2014.

[65] D. R. Hodas, "Appliance Efficiency. UNEP Guide for Energy Efficiency and Renewable Energy Laws (UN Environment Programme, 2016), 17–04.," UN Environment, Nairobi, 2016.

[66] V. Varanavicius, A. Navikaite, Y. Bilan and W. Strielkowski, "Analysis of Consumer Behaviour in Regional Energy Consumption," *Ekonomika regiona [Economy of Region],* vol. 13, no. 1, pp. 147–156, 2017.

[67] J. Munyaneza, M. Wakeel and B. Chen, "Overview of Rwanda Energy Sector: From Energy Shortage to Sufficiency," *Energy Procedia, 104, 215–220,* vol. 104, pp. 215–220, December 2016.

[68] S. Bimenyimana, G. N. Asemota and L. Li, "The state of the power sector in Rwanda: a progressive sector with ambitious targets," *Frontiers in Energy Research,* vol. 6, no. 18, 2018.

[69] J. d. D. Uwisengeyimana, A. Teke and I. Turgay, "Current Overview of Renewable Energy Resources in Rwanda," *Journal of Energy and Natural Resources,* vol. 5, no. 6, pp. 92–97, On 21 January 2017.

[70] T. Uhorakeye, Modelling electricity supply options for Rwanda in the face of climate change, Europa-Universität Flensburg, 2016.

[71] M. Grimm, L. Lenz, P. Jorg and S. Mazimilane, "Demand for Off-Grid Solar Electricity – Experimental Evidence from Rwanda (May 1, 2019)," *USAEE Working Paper No. 19–411,* vol. 19, no. 411, 2019.

[72] G. Li, J. Sun and A. Dai, "Village differences in rural household energy consumption within the Loess hilly region," *Energy, Sustainability and Society,* vol. 6, no. 33, 2016.

[73] N. Mohlakoana, J. N., de Groot, A. Knox and H. Bressers, "Determinants of energy use in the informal food sector," *Development Southern Africa,* vol. 36, no. 4, pp. 476–490, September 2018.

[74] D. T. Manning, P. Means, D. Zimmerle, K. Galvin, J. Loomis and K. Paustin, "Using contingent behavior analysis to measure benefits from rural electrification in developing countries: an example from Rwanda. Energy Policy, 86, 393–401.," *Energy Policy,* vol. 86, pp. 393–401, 2015.

[75] R. Kennedy, S. Numminen, J. Sutherland and J. Urpelainen, "Multilevel customer segmentation for off-grid solar in developing countries: Evidence from solar home systems in Rwanda and Kenya," *Energy,* vol. 186, no. 115728, 2019.

[76] I. Bisaga, N. Puźniak-Holford, A. Grealish, C. Baker-Brian and P. Parikh, "Scalable off-grid energy services enabled by IoT: A case study of BBOXX SMART Solar," *Energy Policy,* vol. 109, pp. 199–207., Octobe 2017.

[77] J. Taminiau, J. P. Banks, D. Bleviss and P. Byrne, "Advancing

transformative sustainability: A comparative analysis of electricity service and supply innovators in the United States," *Wiley Interdisciplinary Reviews: Energy and Environment,* vol. 8, no. 4, 2019.

[78] J. Taminiau and J. Byrne, "Reconsidering growth in the greenhouse: The Sustainable Energy Utility (SEU) as a practical strategy for the twenty-first century. In G. Dale, M. V. Mathai & Puppim de Oliveira, José Antonio (Eds.)," in *Green growth: Ideology, political economy, and the alternatives*, London, Zed Books , 2016, pp. (pp. 233–252).

[79] J. Byrne and J. Taminiau, "A review of sustainable energy utility and energy service utility concepts and applications: realizing ecological and social sustainability with a community utility. Wiley Interdisciplinary Reviews: Energy and Environment," *Wiley Interdisciplinary Reviews: Energy and Environment,* vol. 5, no. 2, pp. 136–154, 2016.

[80] J. Byrne, C. Martinez and C. Ruggero, "Relocating Energy in the Social Commons: Ideas for a Sustainable Energy Utility," *Bulletin of Science, Technology & Society,* vol. 29, no. 2, pp. 81–99, 2009.

[81] C. O'Mara, P. Cherry and D. Hodas, "Clean Energy Policy in Delaware: A Small Wonder. Retrieved September 12, 2020, from http://www.jstor.org/stable/40925249," *Natural Resources & Environment,* vol. 25, no. 2, pp. 3–7, 2010.

[82] I. J. Pérez-Arriaga, R. Stoner, R. Rahnama, S. J. Lee, . G. Jacquot, E. Protzer, A. Gonzalez, A. Reja, M. Brusnahan, P. Duenas and F. J. Santos, "A utility approach to accelerate universal electricity access in less developed countries: A regulatory proposal," *Economics of Energy & Environmental Policy,* vol. 8, no. (1), pp. 33–50., 2018.

[83] A. Faiers, M. Cook and C. Neame, "Towards a contemporary approach for understanding consumer behaviour in the context of domestic energy use," *Energy Policy,* vol. 35, no. 8, pp. 4381–4390, 2007.

[84] J. Barrie and H. J. Cruickshank, "Shedding light on the last mile: A study on the diffusion of Pay As You Go Solar Home Systems in Central East Africa," *IDEAS,* vol. 107(C), pp. 425–436., 2017.

[85] B. Sergi, M. Babcock, N. J. Williams, J. Thornburg, A. Loew and R. E. Ciez, "Institutional influence on power sector investments: A case study of on-and off-grid energy in Kenya and Tanzania," *Energy research & social science,* vol. 41, pp. 59–70, July 2018.

[86] A. Mumma and D. Hodas, "Designing a Global Post-Kyoto Climate Change Protocol that Advances Human Development," *Georgetown International Environmental Law Review (GIELR),* vol. May 20, no. Widener Law School Legal Studies, p. 619, 2008.

[87] A. D. Tigabu, F. Berkhout and P. van Beukering, "The diffusion of a renewable energy technology and innovation system functioning: Comparing bio-digestion in Kenya and Rwanda," *Technological Forecasting and Social Change,* vol. 90, pp. 331–345, 2015.

Chapter 9

Beyond the Hydrocarbon Economy: The Case of Algeria

Cecilia Camporeale, Roberto Del Ciello and Mario Jorizzo

Abstract

The energy sector is vital to efforts to combat climate change as well as achieve economic development. The economy of many Middle East and North African (MENA) countries, such as Algeria, Iran, Qatar, Saudi Arabia, is completely based on hydrocarbons which represent the main source of the state revenue. Investing in renewable energy and efficiency is a winner strategy, allowing both to ensure the necessary availability of energy to cover the country's domestic energy demand and to make more resources available for export to guarantee the state earnings. Renewable sources can be a solution for a transition to a more sustainable economy and a response to the economic stability of these countries affected by the volatility of oil prices. Such a strategy is reflected in improving the attractiveness of foreign investment in the renewable energy sector. Focusing on Algeria, in this article, we analyze the link between the Algerian economy and energy, underlining the current weakness. This work is partially based on the research financed by the meetMED project (WP 3.1) on barriers for domestic and international investors in the energy sector of Algeria.

Keywords: Algeria, barrier to investment, renewable energy, energy efficiency, energy sector, meetMED

1. Introduction

The fossil fuel price fluctuations have large impacts on the economy and often with implications for political stability both for importer countries [1, 2] and for exporter countries, as Middle East and North African (MENA) countries[1], because of its predominance of state revenues. In 2017, the GDP of MENA countries has been more less of 4% [3] of World GDP, while in terms of hydrocarbon production, they accounted the 30% of total world production of oil and natural gas [4]. According to WTO [5], the total merchandise exports of MENA countries are only 6% of total world export, but this amount rises to 26% in terms of world fuel export, with difference among the countries (see **Figure 1**). In Algeria, Iraq, and Kuwait, for example, more than 90% of export is linked to the fuel (respectively, 96%, 94%, and 90%), while in Egypt and in the United Arab Emirates, the percentage falls to 20%, and only in some cases (Israel, Morocco, the Lebanese Republic, and Jordan) the contribution to the fuel export is very residual.

According to Arent et al. [2], many developing countries—particularly MENA countries—may possess an inherent comparative advantage due to the availability

[1] The MENA Countries are composed by Algeria, Bahrain, Egypt, Iran, Iraq, Israel, Jordan, Kuwait, Lebanon, Libya, Morocco, Oman, Palestine, Qatar, Saudi Arabia, Syria, Tunisia, United Arab Emirates and Yemen.

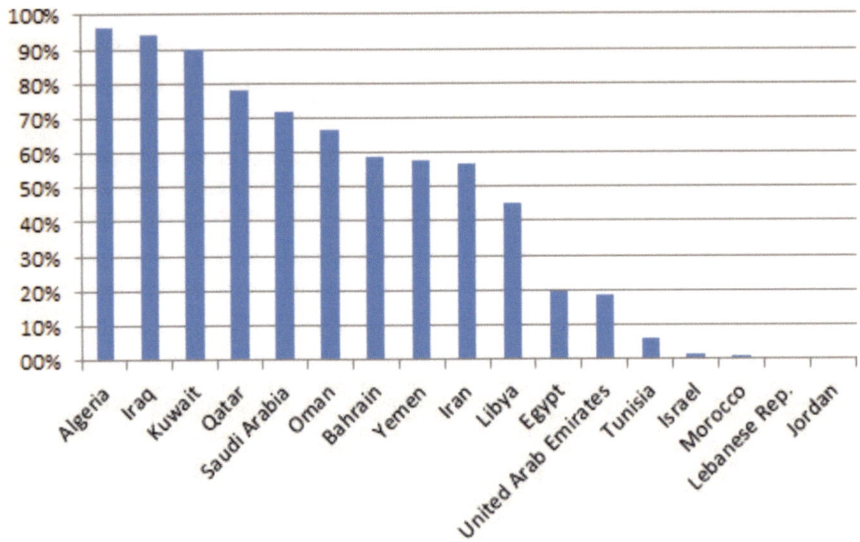

Figure 1.
Percentage of fuel export on total merchandise export in MENA countries, 2017. Source: our elaboration on [5].

of significant renewable energy sources (solar, wind, geothermal, bioenergy, etc.), and taking into consideration the nature of renewable energy particularly relevant for rural and isolated area, it will provide unprecedented opportunity to guarantee electricity access to the poorest citizens. Betting on the renewable sources means obtaining double benefits: on one hand, to assure an economic growth to the citizens [2] and on the other hand, to make fossil resources available for export before to realize a full clean energy transition. Moreover, investing in renewable energy is necessary to achieve the commitments under Paris Agreement on climate change, in general, in order to reduce GHG emissions linked to the use of fossil fuel. This transition toward clean energy boosts the debate on stranded asset, where assets suffer from unanticipated or premature write-offs, downward revaluations, or conversion to liabilities [6], and renewable and energy efficiency investments can be considered a mitigation strategy versus stranded asset risks related to hydrocarbon core business.

A variety of factors could lead to assets becoming stranded: new government regulations that limit the use of fossil fuels (like carbon pricing), boom in more cost-effective (economically, environmentally, and socially) alternatives, a change in demand, and, of course, growing unpopularity with the public opinion [6, 7].

The competitiveness of fossil fuel companies is rapidly losing its attractiveness at an accelerated rate, causing serious problems especially for those countries whose state balance is based on these assets.

In particular, the developing countries are:

- Highly exposed to a decline in fossil fuel demand, consequently to a fall in fossil fuel prices;
- Less able to diversify away from this risk;
- Themselves under pressures to implement policies that may expose them to further risk [8];

The world economy is gradually "decarbonizing" with a continued trend and a deep change in the financial world. Now, many institutions, sovereign wealth funds,

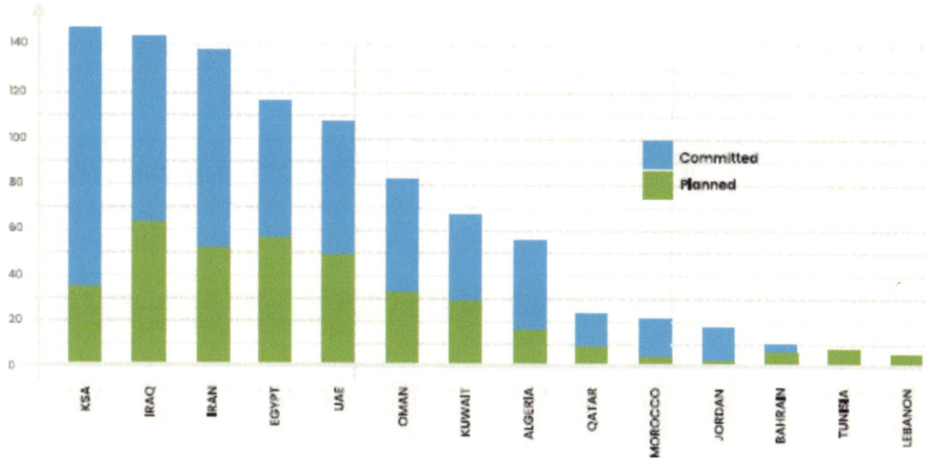

Figure 2.
Total planned and committed MENA energy investment 2019–23 (USD billion). Source: [9].

banks, global asset managers and insurance companies, cities, pension funds, healthcare organizations, universities, faith groups, and foundations have committed to ban fossil fuel investments[2].

According to APIC [9], MENA region will maintain upstream investments with total investment for 2019–2023 near one trillion dollars (see **Figure 2**) even if the 5-year GDP growth forecast has declined. Raising capital is one of the major challenges for regional governments, and recent efforts to attract foreign investment have seen signs of caution from investors, especially taking into consideration the economic performance, solvency, and political reliability of some countries. All these considerations mean that the risk of stranded assets for fossil fuel companies is growing [8].

This work is partially based on the research financed by the meetMED project (WP 3.1) on barriers for domestic and international investors in the energy sector of Algeria.

2. The case study of Algeria

Algeria, officially the People's Democratic Republic of Algeria, is a country in the Maghreb region of North Africa that, with its 2,381,741 square kilometers, is the 11th largest country in the world and the largest in Africa [10].

The country is a semi-presidential republic consisting of 48 provinces and 1541 communes (counties) where the legal system is based on a mix of French civil law and Islamic law. In fact, after more than a century of rule by France, Algerians fought through much of the 1950s to achieve independence in 1962.

Algeria's economy remains dominated by the state, through its state-owned companies which manage the rich fossil fuel resources of the country. Recently, the Algerian government has suspended the process of privatization of state-owned

[2] For example: Norway's sovereign wealth fund, the Catholic Bishops' Conference of the Philippine, the Rockefeller Brothers Fund, the British Medical Association, Amundi Asset Management, Caisse des Dépôts (the French public financial institution), New York City, the City of Cape Town, KfW Group (Germany's development bank), Stockholm University, the Tate museums in the U.K. The National Trust and Allianz insurance, and St Mary's Episcopal Cathedral, Edinburgh—the first cathedral in the world to divest. Recently, also the EIB plans to cut all funding for fossil fuel projects by 2020 in order to focus on long-term investments that must be aligned with the Paris Agreement and, of course, cut greenhouse gas emissions.

industries and companies, imposing restriction on imports and on the participation of foreigners in its economy [10]. In this way, a very explicit import substitution policy has been realized to support the economy of the country and to reduce the uncontrolled exposure to the foreign interference.

The hard core of Algerian economy has long been the hydrocarbons which account for broadly 30% of GDP, 60% of budget revenues, and nearly 95% of export earnings. Hydrocarbon exports enabled Algeria to maintain macroeconomic stability, amass large foreign currency reserves, and maintain low external debt while global oil prices were high.

Since 2014, the Algerian economy has to face the crisis linked to the lower oil prices. In fact, the national foreign exchange reserves were halved and the oil stabilization fund collapsed sharply, reaching its legal minimum value of $7 billion in 2017, compared to a much higher value of $20 billion in 2013, and the large subsidies for the population has fallen under stress [10].

Due to the declining oil prices, the government has been under pressure to reduce spending; therefore, it has reduced its ability to distribute rents and fund generous public subsidies.

3. Economic situation and performance

According to the African Development Bank (AfDB) [11], in many African countries, energy subsidies constitute a considerable fiscal burden, and its contribution to GDP has remained substantially relevant in the time despite the trend of international oil prices. In particular, among oil-exporting economies, there are countries (i.e., Angola, Cameroon, and Nigeria) in which the share of energy subsidies on GDP has the same both in the pre-peak period (2013–2014) and in the post-peak period (2015–2017) but in some others (i.e. Libya, Algeria, and the Congo) where the share increased (**Figure 3**).

The AfDB suggests that subsidy reforms must be geared toward more-efficient and better targeted social safety nets for the most vulnerable. This could improve public finance management, create more fiscal space for much needed public investments in infrastructure, and improve the debt situation.

Over the past 3 years, the Algerian government has enacted incremental increases in some taxes, especially in value-added tax, resulting in modest increases

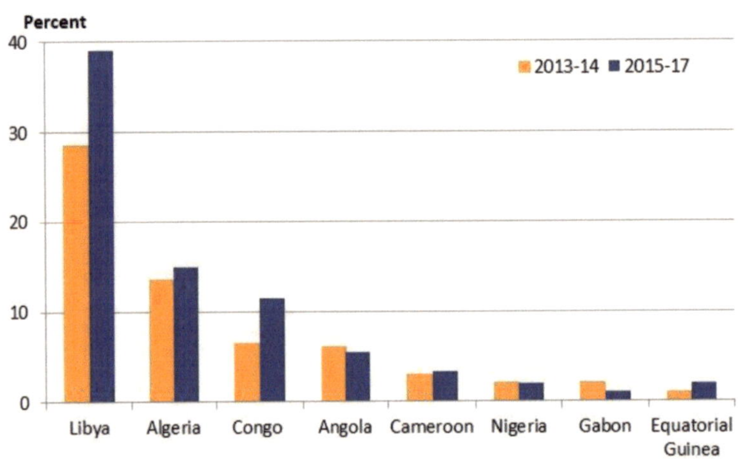

Figure 3.
Energy subsidies as a share of nominal GDP (African oil exporters, 2013–2014 and 2015–2017). Source: [11].

in prices for gasoline and certain imported goods. However, the government is not disposal to reduce subsidies because these subsidies represent measures to assure benefits to poorest people of the country financing some essential services as education, healthcare, and housing programs.

Since 2015, Algeria has strengthened protectionist measures to limit its import bill and encourage greater contribution on GDP of domestic production from non-hydrocarbon industries but also impose additional restrictions on access to foreign exchange for imports.

In 2016, Algerian GDP has increased by 3.8% as the previous year but with different drivers. In fact, the good performance is referred to the recovery in hydrocarbon production, which grew by 3.6% (compared with +0.4% in 2015), more than offsetting the slowdown in growth in the non-hydrocarbon sector, which rose from +5% in 2015 to +3.9 in 2016. The performance of the non-hydrocarbon sector was heavily affected by the slowdown in agriculture due to particularly unfavorable weather conditions, in energy due to the high weight of hidden subsidies, in water, and in other residual industrial sectors [12].

In 2017, the slight decline of hydrocarbon production together with the light growth of non-hydrocarbon sector performance has exerted pressure on the Algerian economy.

Real GDP has grown of only 2.1% rather than 3.3% of the previous year due mainly to the modest production of hydrocarbon sectors. In fact, the hydrocarbon production decreased by 1.4% while the non-hydrocarbon sectors slightly increased to 2.5% (against 2.3% of 2016) due to the inversion of fiscal consolidation happened in second middle of 2017. Inflation remains high although lower than in 2016 (5.5% in 2017 against 6.4% in 2016) [13].

Growth is expected to recover sharply in 2018 as fiscal expansion takes hold. As new public investments announced in the 2018 budget are carried out, headline growth and inflation will increase. As a result, GDP growth will have difficulty to exceed the 2% threshold for 2019–2020, which is a very limited growth for a middle-income country with a high youth population. So, for the next years, many analysts agree to affirm that while the production by new oil wells will boost the economy growth, the progressive increment of contribution to the GDP of non-hydrocarbon sectors could permit to support the fiscal consolidation. In fact, the international oil price fluctuations have underlined the weakness and the fragility of the Algerian economic model and the need to rethinking to a deep structural transformation in which reduce the dependency from hydrocarbon sectors [13].

Public spending decreased by less than expected due to difficulties in pursuing the 2017 budget target. In fact, since the end of the 1990s, Algeria has made massive investments in health and education in response to the pressing needs of its people, while also working to close large infrastructure gaps. In the current fiscal framework (2018–2020), adopted in the 2018 Budget Law, public spending will remain very high and will not be offset by a potential increase of government revenues due to an expected pickup in oil price and production [13].

Public investment has been about 20% of non-hydrocarbon GDP on average since 2000, much larger than in comparator countries. Reflecting the country's policy priorities, Algeria allocated on average about 70% of public investment to economic (i.e., roads, ports, rails, airports, and power and energy) and social infrastructure (such as housing, health, education) [14].

Although imports increased slightly, by 2.7% in 2017, exports have increased significantly, by 16.5%. As a result of continued deficits and limited capital inflows, the country's international reserves declined sharply. Nonetheless, external debt remains very low [13].

In Algerian economy, the role of the state is predominant. Consequently, the public investments are essential to boost the economy. For example, in 2015, the state alone invested the 49% of total gross accumulation of fixed funds. At the same time, the public employment engaged mainly on the labor force in the country: by 2017, the central government employment alone absorbed about 20% of total formal employment. Considering the whole public employment, around 40% of total formal employment is more or less connected to the public staff [14]. The unemployment rate is particularly high (around 12), reflecting the sluggish non-hydrocarbon growth with women participation lower.

4. The challenges of Algerian economy

The Algerian economy is characterized by a strong dependence on the international price of oil, which is the main source of revenue for the state budget. The oil and gas sector is the backbone of the economy, accounting for about 20% of the gross domestic product, and 85% of total exports [15].

Although Algeria remains a relatively closed economy, the government needs to diversify its economy away from hydrocarbons especially since oil prices started falling dramatically in 2014 [16]. So, a range of import restrictions have also been introduced in recent years, as the government attempted to boost domestic production capacity and reduce imbalances in the external accounts.

Oil and gas accounted two thirds of state revenues and, consequently, when financial incentives such as generous subsidies and free housing proved insufficient in stifling popular dissent, these revenues helped the security apparatus acquire the coercive means to repress it [16].

Algeria's high vulnerability to volatile international oil prices exposes it to a high risk of a prolonged economic slowdown. In fact, the predominantly state and nontransparent economic system creates a mistrustful environment for foreign investment, in which state contracts are based on personal knowledge rather than on merit or efficiency, and ends up supporting industries that are not competitive at the international level.

These inefficiencies have also affected the energy sector, which has led to a further economic slowdown. As a result, Algeria became the only OPEC member to pump below the allowed quota as its production decreased, despite efforts to attract new investment.

The budget deficit was significantly reduced in percent of GDP, due to lower spending and higher revenue, but the decline in non-hydrocarbon deficit was more moderate, and deficits were financed largely by drawing on savings in the oil fund, which was depleted in 2017 [17].

In addition to a particularly weak economy, Algeria must also address the threats posed by the climate change that is taking place: most of the country is arid or semiarid; the yearly average rainfall declined by more than 30% over the past decades; more than 50 million of hectares face highly deteriorated conditions; the exodus of rural population toward large cities; and the decline in water resources [18].

So, taking into account the international commitment due to the Paris Agreement and its goals, even underlining limited responsibility in terms of accumulation of GHG as a developing country, the Algerian government declared its willingness to make its contribution as condition of new financial resources, to be combined with traditional partners, and/or of transfer of clean technologies under favorable conditions.

In 2016, the government published details of its "new growth model" to face the significant drop in oil prices, recognizing that its vast system of government subsidies is unsustainable and needs to be revised and the need to boost non-hydrocarbon industries.

Algeria has not financed its deficit through increased external debt, which remains negligible at less than 2% of GDP. Likewise, government debt, consisting mainly of domestic debt, is limited to 40% of GDP.

The volatility of oil prices, the weakness of its economy, the decline in non-oil industrial productivity, and the fueling unemployment bring the government to rethink its vision. This context led authorities in 2016 to adopt the New Economic Growth Model 2016–2030 [19], aimed at structural transformation to reduce the state's role while enhancing that of the private sector and limiting dependency on hydrocarbon revenues. The main reforms relate to improving the business climate and replacing direct and indirect subsidies with targeted social protection for low-income population [11]. All have faced resistance from entrenched economic interests and institutional inertia, which forced them to backpedal on reform [16].

According to the document, subsidies must to be rethinked: in all sectors of the public service (electricity, gas, water, rail transport, telecommunications), tariffs are kept at levels lower than the cost of operations for more than a decade for social considerations.

The document proposes to diversify the economy with a focus on renewable energy, agriculture, and industry. Algeria has long aimed to diversify its largely state-controlled economy, but investment has been hampered by state bureaucracy and inertia.

In addition, the "New Economic Growth Model" aims to promote, among other points, some structural reforms, such as the energy subsidy reform. According to this new vision of economy, the Algerian government, together with the World Bank, has analyzed a strong reform mainly to help vulnerable people of the country. Meanwhile, the government is studying a structural reform plan to modernize the entire system (i.e., simplify business regulations, improve governance and transparency, reform the pension system, and modernize the financial sector). This is a further signal to enhance the country's business climate, which also includes opening up the maritime and airfreight transport sector to the private sector [17].

Algeria aims to reconcile the energy transition and the fight against climate change with the objective of ensuring the right level of well-being for the population, especially the youngest.

5. Energy supply and consumption

Algeria, a country severely affected by desertification, is—like other countries in Africa and in the south of the Mediterranean—particularly vulnerable to the multiform effects of climate change that threaten to undermine its economic and social development [18].

At the same time, the Algerian hydrocarbon energy sector has the main pillar of country's economy both for state balance thanks to export revenues and for the availability of fossil fuel sources.

Algeria's economic performance is highly dependent on the trend in the international price of hydrocarbons, and this has prompted former President Bouteflika to announce—in autumn 2017—Algeria's intention to actively engage in the development of its unconventional energy resources. Algeria has thus committed itself to developing non-hydrocarbon industries, on the one hand tackling the problems associated with heavy regulation and on the other the high emphasis on the growth that the state fears. Algeria has not increased its hydrocarbon exports in this way, but they have decreased due to the progressive exhaustion of deposits and the simultaneous increase in domestic demand [10].

Algeria has the 10th largest reserves of natural gas in the world—including the 3rd largest reserves of shale gas—and is the 9th largest gas producers; it ranks 7th largest gas exporters, 3rd largest gas liquefaction capacity [20, 21], and the 7th largest natural gas liquids producers but also 16th in proven oil reserves [10].

In the recent years, Algeria has beginning to diversify its energy sector through solar energy in order to increase its energy independence. Despite a considerable potential, the share of renewable energies in the energy balance is still low especially in the production of electricity. According to Ghezloun et al. [22], the only condition that the energy mix of Algeria will grow potential of renewable energy is the policy support and encouragement to the introduction of hybrid possibilities, including electricity generation by the private sector.

However, the government aims to develop a photovoltaic industry and, more generally, the renewable energy sector, in order to achieve a win-win strategy in which, on the one hand, a greater share of gas is released for export, thus ensuring the stability of its balance sheet and, on the other hand, the energy necessary for the development of the country's production structure is guaranteed. From the latter point of view, in addition to aiming at a diversification of the source of energy supply, the diversification of the production structure also moves toward the development of sectors and industrial sectors other than the traditional sectors based on hydrocarbons [23].

The energy demand of Algeria is completely covered by its own production, which is almost fully based on fossil fuels. Natural gas is the primary source of power generation contributing to over 93% of installed power capacity. The share of renewable energies in the energy mix is only around 3.4% and until recently was largely dominated by hydropower.

Looking to the Algerian energy flows (**Figure 4**), in 2017, the primary energy production was 165.6 Mtoe with an increasing gas production covering the light decline of liquids (oil and LPG) due to the OPEC agreement of reducing oil supply. The contribution of primary electricity was double compared to 2016 thanks to new renewable energy capacity: 5 new PV plants, equal to a 125 MW total. The natural gas remained the main source followed by oil. The same breakdown is seen for final energy consumption (see **Table 1**), provided by Ministère de l'Energie [24–28] where natural gas remains the main source used to cover 37% of final energy consumption in 2017, followed by electricity (30%).

Algeria is moving toward renewable energy direction, having announced a substantial 20-year plan for solar development, which calls for 5% renewable energy installed capacity by 2017, and 20% by 2030, of which 70% would be CSP, 20% PV, and the remaining 10% wind power [23]. Hosting one of the world's first Integrated Solar Combined Cycle (ISCC) plants (the ISCC Hassi R'mel is a 150-MWe combined cycle hybridized with a 25-MWe equivalent CSP solar field. It was the first ISCC plant in the world to start construction although Morocco's ISCC Ain Beni Mathar was the first operating plant of this type in the world [29]), Algeria also has gained a valuable insight into the development, construction, and operation of this type of plant.

Approximately 6.2% of gas production was oriented to the production of electricity (17.5 Mtoe vs 16.5 Mtoe in 2016).

The export volume was equal to 108.3 Mtoe, in reduction of −2.2% compared with 2016. This decline affected almost all products except LNG and electricity, which recorded increases of 5.7% and 71.2%, respectively. Imports, 4.2 Mtoe, were mainly linked to a derived energy (3.9 Mtoe) due to imports of coke and electricity, which offset the decline in imports of petroleum products (−3.4%), but while remaining a net exporter (104.1 Mtoe).

National final energy consumption (including losses) reached 59.6 Mtoe in 2017, mainly driven by final consumption (+4.1%). Conversely, non-energy

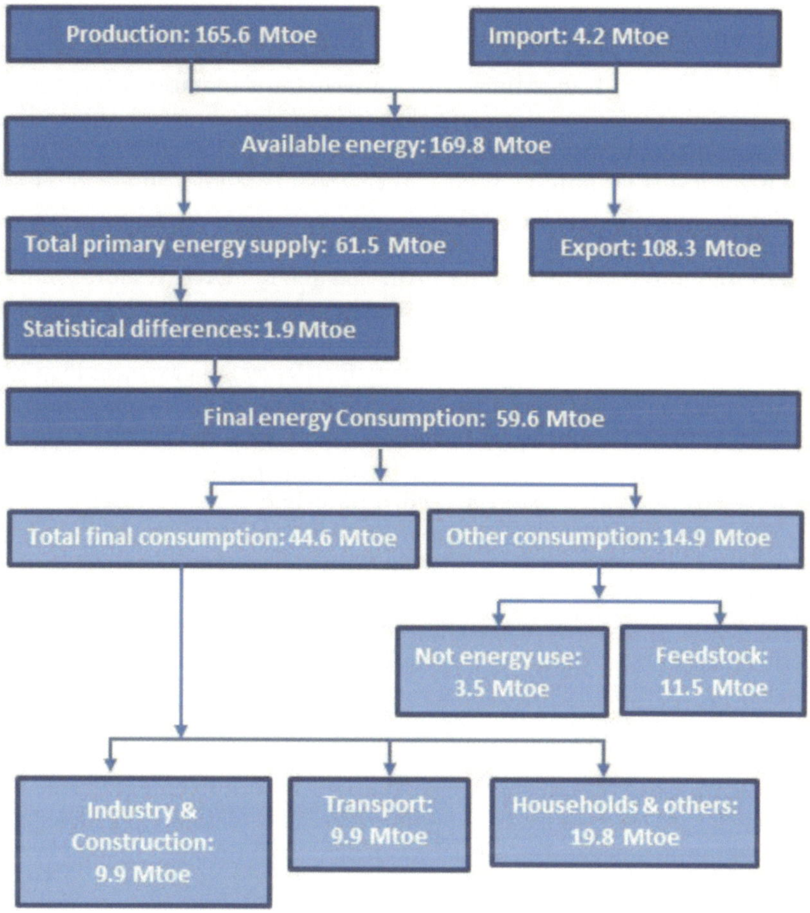

Figure 4.
Synthesis of energy flows (Mtoe) in 2017. Source: [28].

Mtoe	2005	2010	2015	2016	2017
Natural gas	13.14	14.46	21.35	21.73	22.03
Oil and oil products	10.54	13.27	17.88	17.18	16.97
Electricity	9.75	12.20	16.41	16.88	17.81
Others	2.83	2.68	2.63	2.55	2.77
Total energy consumption	36.19	43.82	58.27	58.34	59.58

Source: Elaboration on [24–28].

Table 1.
Algeria: national final energy consumption by fuels.

consumption, which refers to the quantities consumed as raw material in the petrochemical and other industries, fell by −19.5%, followed by the consumption reduction of energy industries (−5.1%).

5.1 Current energy plan of the country

The general approach and strategy intended by the Algerian Ministry of Energy and Mines is constituted by decree no. 07-266, dating on 9 September 2007 describing the function and role of the Ministry with respect to the intentions of the Algerian government.

In particular, in Art. 1, the Ministry commits to the elaboration of political and strategic research; the production and valorization of hydrocarbon, mineral, and energetic resources; and the embedment of the respective industry in this sector, while in Art. 5 it furthermore commits to the necessary studies and research and the promotion of sources of renewable energy.

Algeria has embarked on renewable energies in order to provide both global and sustainable solutions to environmental challenges and fossil fuel resource conservation. To achieve these two targets, Algeria has launched an ambitious programme for the development of renewable energy which was adopted by the Government in February 2011 (Renewable Energy and Energy Efficiency Development Plan 2011–2030) and revised in May 2015 in terms of some adjustment of the renewable targets.

Some energy policies to support the implementation of renewable energy sources were made since 2004 (Law on Renewable Energy Promotion, see **Table 2**, data provided by IEA [30]), before as studies of technologies (research program) and regulatory instruments than as economic instrument to promote renewable energy resource through incentives (feed-in tariffs, premiums, or direct incentives).

According to the revised strategy, new ambitious program for the development of renewable energy 2015–2030 aims to reach a contribution of renewable energy sources in term of power capacity of 37% (22 GW) by 2030 with more than 4.5 GW to be realized before 2020. The share of renewable energy in electricity generation should thereby reach 27% (previously 20%) by 2030. These targets have been included in the nationally determined contribution, which Algeria has sent to the UNFCCC secretariat as a contribution to attain the Paris Agreement (**Table 3**).

To make this, Algeria should be supported by CTCN (Climate Technology Centre and Network) intervention. The CTCN will help the photovoltaic market in Algeria with a specific project on the design and construction of a ground base's 1 MW photovoltaic plant and with a project still in its implementation phase focused on the establishment of a laboratory for accreditation and quality control of photovoltaic

Title	Year	Policy type
Renewable Energy and Energy Efficiency Development Plan 2015–2030	2015	*Policy support*: strategic planning
Feed-in tariff for solar PV installations	Apr 2014	*Economic instruments*: fiscal/financial incentives
Renewable Energy and Energy Efficiency Development Plan 2011–2030	Feb 2011	*Policy support*: strategic planning
Renewable Energy National Fund	2009	*Policy support*: institutional creation. *Economic instruments*: fiscal/financial incentives
Law 04-92 on the Diversification of Power Generation Costs (REFIT)	2004	*Economic instruments*: fiscal/financial incentives
Law 04-90 on Renewable Energy Promotion in the Framework of Sustainable Development	2004	*Regulatory instruments*: codes and standards *Policy support*: institutional creation *Research, development, and deployment (RD&D)*: research program, technology deployment, and diffusion
Law 99-09 on the Management of Energy	1999	*Policy support*: strategic planning, institutional creation

Source: Elaboration on [30].

Table 2.
Algerian energy policy in force.

	1° step: 2015–2020	2° step: 2021–2030	Total
Photovoltaic	3000	10,575	13,575
Wind	1010	4000	5010
CSP	-	2000	2000
Cogeneration	150	250	400
Biomass	360	640	1000
Geothermal	5	10	15
Total	4525	17,475	22,000

Table 3.
Algerian renewable energy programme 2015–2030 (MW).

modules [31, 32]. In this way, Algeria could achieve its renewable energy target, reinforce the national know-how, and build specific competence in the sector.

6. The foreign investor flows

Since 2008, foreign investors have been restricted to a maximum stake of 49% in a company. This policy has, more probably, the reason because the foreign direct investment (FDI) flows to Algeria have diminished in recent years.

In 2015, Algeria registered its first negative foreign trade balance since 1994, and despite falling revenues, Algeria—which has the largest defense budget in Africa—opted not to cut military spending.

In 2016, Algeria launched a new operation of financing economic investments and major infrastructure projects called "National Bond Issue for Economic Growth," raising $5.2 billion to finance its domestic debt market. Previously reluctant to borrow on international markets, the government obtained a $1 billion loan from the African Development Bank [33]. Through this instrument, the Algerian Ministry of Finance invited his nationals living abroad to subscribe to this national bond issue, open to public subscription since April 2016, and participate in the economic development of the country.

However, according to the United Nations Conference on Trade and Development (UNCTAD) [34], the foreign investments in Algeria fell slightly in 2016–2017, due to the strong dependence of Algerian economy by fossil fuels in which prices fell in 2017, but they should be recovered in the next years thanks to the reform on investment laws, proved by the heavy investments made by China and Turkey in this last years.

The main obstacles to the boost of investments in the country are identified in the protectionism measures, as well as corruption, bureaucracy, a weak financial sector, and legal insecurity in terms of intellectual property rights.

According to IMF [35], fiscal risks in Algeria are multiple and interrelated due to the dominant role played by the state in economic activity through government programs as well as through commercial activities carried out by public institutions and state-owned enterprises.

Other sources of fiscal risk include volatile hydrocarbon revenues, natural disasters, and the financial situation of social safety net programs. According to some estimates, in 2016, economic losses linked to the country's climate, which was not fully conducive to investment, were quantified at around 8.9% of GDP. This loss reflects, on the one hand, the government's purchase of the debt of a public service company held vis-à-vis a bank that is also public and, on the other hand, the issue of bonds by the state-owned oil company to compensate for losses resulting from the sale of imported refined fuels on the domestic market at preferential prices [35].

However, Algeria is seeking more trade and foreign investment. For example, in April 2005, the hydrocarbons law was designed to encourage foreign investment in energy exploration, or in 2016 the "revised investment law" (Law 16-09) and the 2016 Finance Law aimed to replace most provisions of the current investment legislation. The Revised Investment Law of 2016 eased restrictions on transferring invested capital, dividends, and disposal proceeds out of the country.

The main challenge of the new framework is to remove the obligation for foreign investors to generate a foreign exchange surplus for the benefit of Algeria over the lifetime of a given investment. In practice, before the new investment framework, foreign investors were constrained to repatriate dividends and profits from their activities in Algeria. Today, with the new reform, on condition of a corporate structure in which the foreign investors are co-owners of an Algerian company, they are free to repatriate the profits that they obtain from such an investment. The clear objective of this reform is, therefore, to make the country more attractive, thus stimulating and gathering more foreign investment.

In addition, the Algerian government is working on a draft law that could also abolish the requirement to involve local partners in the participation of foreign contractors in public tenders; the draft law is expected to be drafted in later 2019. These are measures to remove the protectionist regime that has characterized Algeria up to now.

Although the investments in Algeria are complex, those in renewable energy are highly risky both in global level and in the Algerian context.

7. Conclusion

In 2014, for the first time in history, the amount of new renewable generation capacity surpassed that of new fossil fuel-based systems on a global basis [36].

The availability of technology is not in itself sufficient to accelerate a change in energy system to assure a clean energy transition; public policy and regulation, market reforms, private sector engagement, stranded asset risk management, and strong analytical tools and data remain important factors [2].

As hydrocarbon revenues in MENA countries make up a significant portion of the government budget and contribute greatly to GDP, the fluctuation of oil price, the need to diversify revenues compared to oil and gas exports, and the need to assure energy for the country's development should expose them to a deep economic and financial crisis.

Generally, the four main factors to decide the investment in renewable energy sources are the following: (i) production, productivity and costs of production factors; (ii) demand, expected internal and external demand for solar components; (iii) risk and stability, real and perceived risks; and (iv) business support, specific support and enabling environment [23].

According to Watts [37], the perception of financial risk is particularly significant in renewable energy projects because they are often capital-intensive and are typically highly leveraged, with up to 70–80% of the project total being financed through debt, but nevertheless there are the possibility to manage the risks by means of risk mitigation and risk transfer. In this way, it is possible to overcome the political and regulatory risks, weather-related volume risk, and other risks.

The difficulties of making an investment, in particular an investment in renewable sources, are becoming particularly acute in Algeria. In fact, focusing on renewable sources would allow the country, on the one hand, to have a cleaner, cheaper, and more sustainable energy system, be able to meet the challenges of a growing energy demand, and, on the other hand, be able to allocate more resources to exports.

However, investing in renewables, in a country with an economy so strongly linked to fossil fuels, is highly challenging.

Although the country can count on two major national competitors with over 20 years of experience in the field of energy, it is true that it is mainly linked to the exploration, development, and exploitation of renewable sources.

The Algerian government has launched several projects to increase production from renewable sources, but precisely because of the investments and the change of pace that they require, they are still limited despite their strong potential.

Local companies have difficulties in making high-calibre investments and, above all, lack the technical knowledge to do so.

The objective of attracting foreign investments to the country is highly resisted by the strong presence of the state in the economy, only marginally encouraged by a large openness abroad.

However, it will be necessary to wait until the next few years before the first effects on the country's economy might be felt through a large opening up to investment in low-carbon technologies and understand whether the path towards sustainable development has really been taken.

Notes

The document is the first step of a working progress country report that will be carried out within the framework of WP 3.1 of the wider MeetMED project in which it is intended to analyze, also through specific interviews (step 2), the main barriers that domestic and international operators face to invest in a renewable energy sources and energy efficiency in Algeria.

In this first step, we have collected the main information relating to the country, analyzing the context in which we find ourselves operating, with particular emphasis on the strong link between economy-state and fossil sources.

In fact, the project foresees in a second step the ad hoc realization of special interviews with institutions, local companies and foreign companies, in order to identify the elements that most block the realization of such investments.

The aim is to produce a country report that can suggest possible lines of action to the policy maker to improve the confidence and attractiveness of the investments, given the economic convenience.

Author details

Cecilia Camporeale*, Roberto Del Ciello and Mario Jorizzo
ENEA SSPT, C.R. Casaccia, Roma, Italy

*Address all correspondence to: cecilia.camporeale@enea.it

IntechOpen

© 2020 The Author(s). Licensee IntechOpen. Distributed under the terms of the Creative Commons Attribution - NonCommercial 4.0 License (https://creativecommons.org/licenses/by-nc/4.0/), which permits use, distribution and reproduction for non-commercial purposes, provided the original is properly cited. BY-NC

References

[1] Arndt C, Hussain MA, Jones ES, Nhate V, Tarp F, Thurlow J. Explaining the evolution of poverty: The case of Mozambique. American Journal of Agricultural Economics. 2012;**94**(4):854-872

[2] Arent D, Arndt C, Miller M, Tarp F, Zinaman O. The Political Economy of Clean Energy Transitions, A study prepared by the United Nations University World Institute for Development Economics Research (UNU-WIDER). Oxford University Press; 2017. DOI: 10.1093/oso/9780198802242.001.0001

[3] World Bank Databank. Available from: https://databank.worldbank.org/home

[4] BP. Statistical Review of World Energy 2019. 2019. Available from: https://www.bp.com/content/dam/bp/business-sites/en/global/corporate/pdfs/energy-economics/statistical-review/bp-stats-review-2019-full-report.pdf

[5] WTO Data. Available from: https://data.wto.org/

[6] Ansar A, Caldecott BL, Tilbury J. Stranded Assets and The Fossil Fuel Divestment Campaign: What Does Divestment Mean for the Valuation of Fossil Fuel Assets? Smith School of Enterprise and the Environment, University of Oxford; 2013

[7] Generation Foundation. Stranded Carbon Assets Why and How Carbon Risks Should Be Incorporated in Investment Analysis. Generation Foundation; 2013. Available from: https://www.genfound.org/media/1374/pdf-generation-foundation-stranded-carbon-assets-v1.pdf

[8] Manley D, Cust J, Cecchinato G. Stranded nations? The climate policy implications for fossil fuel-rich developing countries. In: OxCarre Policy Paper 34. Department of Economics, Oxford Centre for the Analysis of Resource Rich Economies (OxCarre), University of Oxford; 2017

[9] APIC. MENA Annual Energy Investment Outlook 2019. Arab Petroleum Investments Corporation; April 2019

[10] CIA. Country Comparison: Area. CIA World Factbook; 20 March 2019. 2018

[11] AfDB. African Economic Outlook (AEO). African Development Bank; 2019. Available from: https://www.afdb.org/en/countries/north-africa/algeria/algeria-economic-outlook/

[12] World Bank. Algeria's Economic Outlook—April 2017. 2018

[13] World Bank. Algeria's Economic Outlook—April 2018. 2019

[14] IMF. IMF Country Report No. 18/169. Algeria: International Monetary Fund; 2018

[15] OPEC. Algeria facts and figures. 2018. Available from: https://www.opec.org/opec_web/en/about_us/146.htm

[16] International Crisis Group. Breaking Algeria's Economic Paralysis, Middle East and North Africa Report N. 192. Released 19 November 2018. 2018

[17] IMF. Algeria—Staff Report for the 2018 Article IV Consultation. International Monetary Fund; 2018

[18] République Algérienne Démocratique et Populaire. Contribution Prévue Déterminée au niveau National CPDN—ALGERIE (called "INDC-Algeria"); 2015

[19] Ministère des Finances. Le Nouveau Modele de Croissance—Synthese. République Algérienne Démocratique et Populaire. 2016

[20] ENI. World Oil Review 2018—Vol. 1, Scenarios. Strategic Options and Investor Relations Department; 2018

[21] ENI. World Gas and Renewables Review 2018—Vol. 2, Scenarios. Strategic Options and Investor Relations Department; 2018

[22] Ghezloun A, Chergui S, Oucher N. Algerian energy strategy in the context of sustainable development (legal framework). Energy Procedia. 2011;**6**:319-324

[23] International Bank for Reconstruction and Development/The World Bank. Competitiveness Assessment of MENA Countries to Develop a Local Solar Industry. MENA ENERGY SERIES—REPORT NO. 94834-MNA; 2015

[24] Ministère de l'Energie et des Mines. Bilan Energétique National de l'année 2005. Edition 2006

[25] Ministère de l'Energie et des Mines. Bilan Energétique National de l'année 2010. Edition 2011

[26] Ministère de l'Energie. Bilan Energétique National—Année 2015. Edition 2016

[27] Ministère de l'Energie. Bilan Energétique National—Année 2016. Edition 2017

[28] Ministère de l'Energie. Bilan Energétique National—Année 2017. Edition 2018

[29] National Renewable Energy Laboratory (NREL). Concentrating Solar Power Projects by Technology: Parabolic Trough Projects. 2013. [Online]. Available from: www.nrel.gov/csp/solarpaces/parabolic_trough.cfm

[30] IEA/OECD/IRENA. Policy and measures. 2018. Available from: https://www.iea.org/policiesandmeasures/renewableenergy/?country=ALGERIA

[31] Grin J, Rotmans J, Schot J. Transition to sustainable development: New direction in the study of long term transformative change. In: Routledge Study in Sustainability Transformation. NY/London: Routledge; 2010

[32] Kemp R. Technology and the transition to environmental sustainability: The problem of technology regime shift. Futures. 1994;**26**:1023-1046

[33] Pera JR. Transformation of Algeria's oil-based Economy is unlikely. IDA Africa Watch. 2017;**15**

[34] UNCTA. World Investment Report 2018. 2018. Available from: https://worldinvestmentreport.unctad.org/world-investment-report-2018/#key-messages

[35] IMF. Algeria—Staff Report for the 2017 Article IV consultation. In: IMF Country Report No. 17/141. International Monetary Fund; 2017

[36] Sawin JL, Sverrisson F, Rickerson W, Lins C, Williamson LE, Adib R, et al. Renewables 2015 Global Status Report: Annual Reporting on Renewables: Ten Years of Excellence. Paris: REN Secretariat; 2015

[37] Watts C. Managing the Risk in Renewable Energy. Swiss Re: The Economist Intelligence Unit Limited; 2011

Chapter 10

Remotely Sensed Data for Assessment of Land Degradation Aspects, Emphases on Egyptian Case Studies

Abd-alla Gad

Abstract

Remote sensing and thematic data were used to provide comprehensive views of surface conditions related to land degradation and desertification, considered environmental extremes in arid and semi-arid regions. The current work applies techniques, starting with simple visual analyses up to a parametric methodology, adopted from the FAO/UNEP and UNESCO provisional methodology for assessment and mapping of soil degradation. Egyptian case studies are highlighted to insinuate on studied aspects. Variable satellite imageries (MSS, TM, and ETM) and aerial photographs were utilized to provide data on soil conditions, land cover, and land use. IDRISI and ArcGIS software were used to manage thematic data, while ERDAS IMAGIN was used to process satellite data and to derive the normalized difference vegetation index (NDVI) values. A GIS model was established to modify the universal soil loss equation (USLE) calculating the present state and risk of soil degradation. The study area is found exposed to slight hazard of water erosion, however, and to high risk of wind erosion. It is also threatened by a slight to high salinization and slight to moderate physical degradation. It is recommended to use a GIS in detailed and very detailed studies for evaluating soil potentiality in agricultural expansion areas.

Keywords: soil degradation, desertification, extreme environment, arid, remote sensing, GIS, Egypt, Sinai

1. Introduction

An extreme environment is a habitat characterized by harsh environmental conditions, beyond the optimal range for the development of humans [1]. For an area to be considered an extreme environment, it must contain certain conditions and aspects that are considered very hard for other life forms to survive.

It was recently realized [2] that numerous political decisions linked with land degradation and desertification research findings were promoted. Examples of such support are referred to the sustainable development call, issued at the UN Conference on Environment and Development. Also, the threat recognition to human welfare through processes of desertification represents that support expressed in

IntechOpen

the United Nations Convention to Combat Desertification [3]. In this context, it has also been emphasized that research should support policy makers and administrative authorities dedicated to establishing locally adapted schemes for sustainable land management.

Although the methods for large scale land degradation assessments are commonly employed and well developed, a monitoring of land degradation and desertification processes, at more detailed, spatial and temporal scales using remote sensing data, is still an issue. This problem is documented in numerous publications discussing the severity of the space images that offer a synoptic view, multispectral and multi-temporal possibilities, and are nearly orthogonal. The synoptic view is possible because about 3.5 million hectares (8.6 million acres) of the earth's surface can be examined on each scene, and all the objects can be compared across the entire scene [4]. The multispectral capability of the satellite images allows the establishment of unique spectral signatures for vegetation and soil related objects. The temporal capability permits the soil test of soils, vegetation, and atmosphere at intervals of different periods (i.e., 18 days for the Landsat images and 1 day for the Meteosat). Due to the very high sensing altitude, the metric distortions of the image are very small. It is thus possible to use the same scale of the map and via superposition of images, with only minor differences.

Exploiting the information provided by remote sensing, it is essential to consider that, it is mandatory to provide stakeholders with an attractive scheme assessing present resources and, if possible, the temporal progress thereof. This objective cannot be met by conventional approaches alone, which commonly rely on field based mapping of ecological parameters, providing a high level of details but only limited spatial coverage. Further information is needed for larger landscape units (e.g., through integrating field-based approaches with remote sensing techniques). Methodologies based on remote sensing data make use of the synoptic advantage, repetitive coverage, and consistent perspective over large areas. Combined with the functionality of today's GIS generation, a powerful tool is provided for monitoring and assessing areas under the threat of the extremes, as land degradation and desertification.

2. Basic theories of land degradation and desertification monitoring

A weakness of many approaches analyzing land degradation and desertification processes is the lack of an intangible outline, defining the reasons for undertaking monitoring in a certain way and how related difficulties could be faced. The necessary outline is not only restricted on accurate description of the respective threats but also a definition of the suitable methodology to face these challenges [5]. On the one hand, remote sensing data play an important role as one of the major sources of up-to-date and physically based information. On the other hand, GIS delivers the toolbox that enables data integration, analysis, and information extraction.

2.1 Processes, indicators, and scale

In order to understand how remote sensing and GIS may support the evaluation of land degradation/desertification threats for different ecosystems, it is needed to understand the environmental setting of these systems. Only a precise knowledge about governing processes leads to monitoring options and to develop meaningful conclusions on how to combat the respective threats. There have been numerous

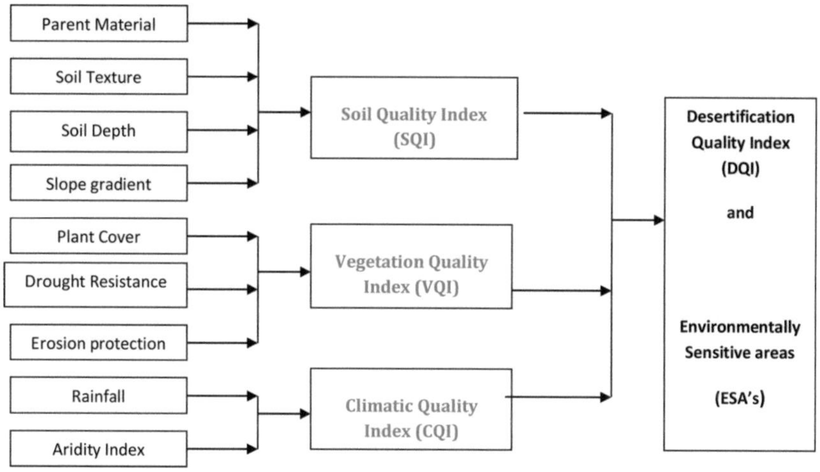

Figure 1.
Flow chart of mapping environmentally sensitive areas (ESAs).

approaches to conceptualize a chain from triggering processes to potential effects. In the current chapter, both visual analyses and digital image processing develop a concept on how to derive useful indicators for land degradation/desertification monitoring on the basis of examples from the European Mediterranean.

2.2 Land degradation/desertification assessment methodologies

A first and simple operated approach may concentrate on geological/geomorphological, climatic, and anthropogenic factors and their interrelationships. This basic approach has been expressed in the assessment of the present status, rate, and risk of land degradation processes [6]. It is included in EU funded DISMED project for assessment and mapping environmentally sensitive areas—**Figure 1**. Land degradation and desertification risk assessment depend upon factors determining changes in soil or vegetation properties. While we have to understand the logical chain of` determinants, we have also to derive pathways on how to conclude from those processes on relevant indicators.

3. Environmental extremes in Egypt

3.1 General

Depending on the extreme physicochemical conditions that characterize the extreme environments, they are classified as *extreme temperature*: extremely cold environments and extremely hot environments, extreme pH: extreme acidic and extreme alkaline environments and *extreme pressure* environments are those under extreme hydrostatic or lithe pressure, such as aquatic habitats at depths of 2000 m or more or deep-subsurface ecosystems [1].

A region is arid when it is characterized by a severe lack of available water, to the extent of hindering or preventing the growth and development of plant and animal life [7]. Environments subject to arid climates tend to lack vegetation and are called xeric or desert. As located in the arid region, over two-thirds of Egypt is covered by the Western Desert. The desert is always dry, but it is actually not lacking in water. Occasional rains fill huge underground aquifers beneath the desert, which occasionally break through to the surface. Just like along the Nile, water is the key to life here and where the water break through the surface thriving oases have formed around the springs. These isolated gardens in the desert have long supported substantial communities of people and substantial agricultural development as well as a culture unique from that of the Nile Valley.

3.2 Assessment of wind and fluvial action

Wind and fluvial actions have been fighting a wining battle against the small strip of the fertile Nile Valley. Historically, the western desert was a grave of Cambyses, king of Persia (50,000 soldiers) when he tried to attack the Siwa oases, Egypt (**Figure 2** shows the oases geographic distribution). The history also recorded about the lost oases in Wadi Hens, between Baharia oases north and Farafra oases in the south. These oases were covered by Eolian deposits. On the western side of the Nile Valley, the agricultural land in oases and depressions (**Figure 3**) is most suffering from the wind action. At the Kharga oases, dunes submerge roads and houses and encroach upon fields and whole villages. A complete village "Ginah" has been engulfed by sand in 1971. The government had built a new Ginah at short distance away to resettle farmers, however, is threatened now by other dune belts.

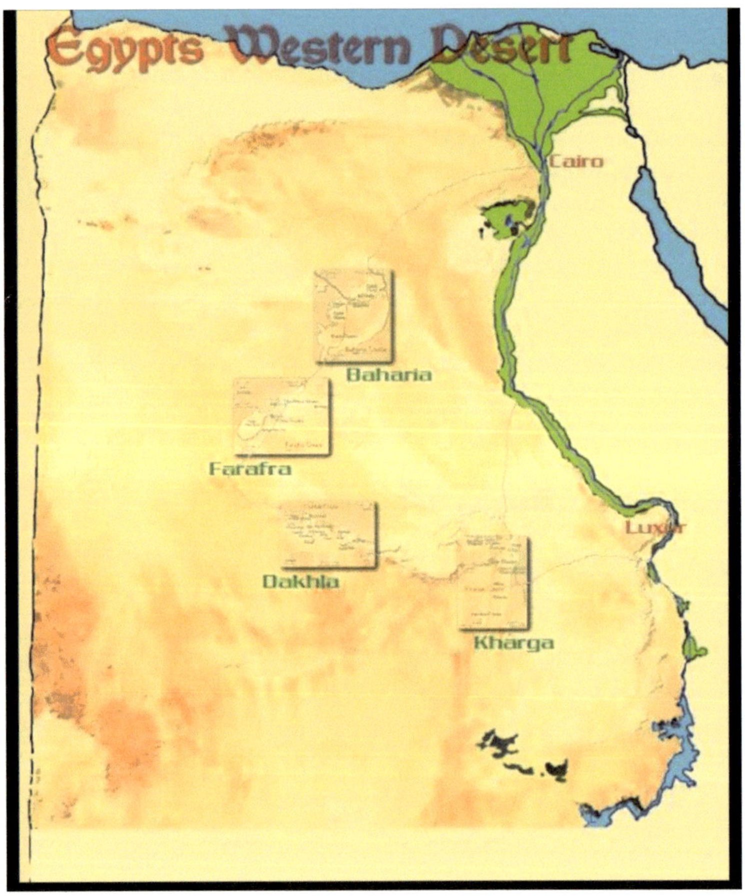

Figure 2.
Geographic distribution of western desert oases, Egypt.

Figure 3.
Eolian deposits were blowing away, leaving deflation residuals in depressions, Egypt.

Despite the fact that methods for large scale degradation/desertification assessments are widely employed and well developed, however, monitoring at appropriate spatial and temporal scales and with adequate remote sensing methodologies is still uncommon. This problem is well documented in numerous publications discussing the severity of land degradation/desertification impacts from regional to global scale (e.g., [8–11]). Accordingly, impact assessments, of land degradation and desertification necessitate a thorough quantification of indicators in the context of global change research, as these processes inherently feedback into the integral development of global economy [7].

On the other side of the valley, the agricultural land is most suffering from the fluvial action. The heavy rainfall during thunderstorms causes severe erosion, threatening the cultivated land. The Menia Governorates has been exposed to strong destructive thunderstorms in the period between 1965 and 1975. The great amounts of the flooding water with their great power and suspended material have caused great physical damages in many villages with the consequent social extreme results. Also, a great thunderstorm has caused a great damage in the Qena area on April 13, 1985. One of the famous torrential extremes is one fallen over Nowaiba, on Red Sea coast in 1988, causing hundreds of deaths and great damages.

The current study aims to highlight the role of remote sensing techniques in assessment of wind and fluvial actions.

3.3 Detection of wind action

A simple visual interpretation of a false color composites (FCCs) of an ETM Landsat mosaic of 2014, covering Egyptian territories for bands 1, 2, and 3 rendered, respectively, in yellow, cyan, and magenta, was used. The images are enhanced photographically to get a maximum contrast in the desert area (**Figure 4**). The images revealed the Eolian deposits in the western desert in bright colors with some inclusion of dark patches. The rock land appears in dark colors mixed in some places with patches of bright ones indicating the existence of Eolian deposits and the rock land appear in different shapes according to their nature.

The satellite images show sand sheets located at the west of Nile Delta and along the western borders of the Nile Valley. The image is characterized by alternating light and dark streaks. The field investigation showed that these areas are covered by extensive sand sheets on an undulating land scape. Some areas were found to be

The image is used for delineation of different physiographic units ;

1) *Undulating landscape "Teraces".*
2) *undulating landscape "Fan system"*
3) *Wadi bottoms*
4) *Windblown soils*
5) *Rock land*
6) *Denuded rock land*
7) *Alluvial Nile plain*
8) *Water body*

Figure 4.
Color composite of ETM (2014) bands 1, 2, and 3 rendered, respectively, in yellow, cyan, and magenta.

covered by gravel and pebbles rather than by sand. It was found that the deposits are characterized by the alternation of layers (1–2 cm) of medium to fine sand and gravel. The image in the FCCs can be interpreted as, whereas light streaks are visible, as areas of deposition of fine-sized material (sand), the dark streaks are areas of non-deposition, erosion or bedrock surfaces.

3.4 Sand dune belts

These Eolian features can be observed developed out of denude able rock and at the south of the Faiyum Depression and extend in an elongated belt parallels to the Nile Valley. They are characterized by light colors, in the shape of parallel strips, alternating with dark colored strips in the same shape (**Figure 5**). The enlargement of Landsat band 3, to scale of 1:100,000 (**Figure 6**) reveals the individual existence of longitudinal dunes, inter-dune areas, and barchans dunes. The density of the longitudinal dunes differs: denser in the south of the Faiyum Depression while it is less dense in the west of the Menia. This variation in dune density is attributed to topographic effects, where scarps exist and increase in height, and they form a barrier to the prevailing winds that may enhance the relative effectiveness of winds from other directions. The longitudinal dune orientation indicates that the main wind direction is N-NW to S-SE.

Barchanoid dunes appear in the same Landsat ETM, as linear ones attaining barchans shaped ridges. Individual barchans, appearing in bright colored rounded and crescentic shaped patches are generated during extreme sand storm events.

Figure 5.
ETM (2014) Eolian features developed out of denude able rock south of the Faiyum Depression and extend in an elongated belt parallels to the Nile Valley.

Figure 6.
The enlargement of Landsat 7, band 3, 2014 to scale of 1:100,000 reveals the individual existence of longitudinal dunes, inter-dune areas, and barchans dunes.

They exist in the leeward of some individual dunes. Some barchans dunes are large enough to indicate a general north-south direction with some deviations, having a west-east direction. Between the sand dunes, dark streaks occur in straight parallel orientation. These are indicating the gravelly corridors of the inter-dune areas, including faint appearing hydrographic network extending from the denuded rocky plateau to the accumulation areas of the debris material in the dune field, as a result of torrential action and the occasional thunder storms. These deposits are reworked by the wind during the long dry periods.

3.5 Erosional patterns

The synoptic view of Landsat images allows to see the rocky obstructs and their influence on the distribution of windblown sands. Isolated hills are also visible as obstacles to the wind transported sand. The rocky terrain appears on the FCCs in dark colors. Different color shades are attributed to importance of water bedrock outcrops and to the degree of erosion. Dark colors exist combined with light colors of the Eolian deposits. These are attributed to the denuded rock land, upon which windblown sands are deposited.

3.6 Detection of fluvial

The fluvial action is most clear in the eastern desert. It is possible to follow (**Figure** 7) the dense dendritic network of the wadis and ravines, which both are characterized by light colors, fine image texture, and dendritic pattern. These fluvial land forms are situated in the high plateaus. Areas of debris accumulation are distinct by their light color patches, which are situated adjacent to the east of the Nile Valley (**Figure 8**). Also, wide wadis are indicating the existence of debris material. In the wadi bottoms, huge amounts of material are deposited. The deposits are reflected in light colors controlled by wadi configuration. Wadi El-Bustan in the south of the eastern desert is a typical example. Most probably, the pattern of different wadis and tributaries was controlled by the area of weakness, and the run-off water of thunderstorms will follow these weak zones. The debris material endangers the eastern cultivated strip of the Nile Valley.

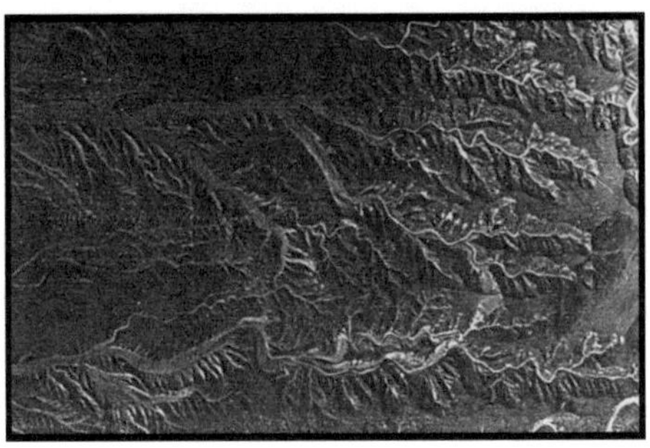

Figure 7.
Landsat ETM7—band 2 (2014) scale of 1:100,000 reveals fluvial degradation land forms at the Easter desert, Egypt, including dendritic drainages, ravines, and gullies.

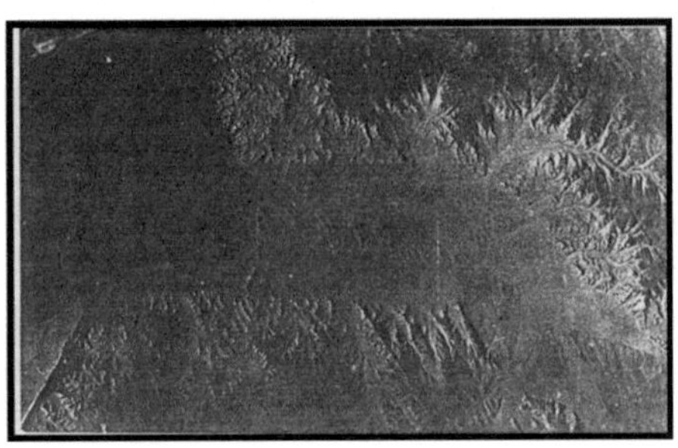

Figure 8.
Landsat ETM7—band 3 (2014) scale of 1:100,000 reveals fluvial depositional land forms at the Easter desert, Egypt, including alluvial fans.

4. Using Meteosat in the study of a thunderstorm on the eastern desert, Egypt

Meteosat sensors have been used for many meteorological applications [12, 13]. It was used for example in wind victor determination [14]. Meteosat data have also been used in the monitoring of natural disasters [15]. Gombeer [16] has used Meteosat and GOES (Geostationary operational Environmental satellites) images of West Africa to compare the corresponding areas on the 1:5,000,000 FAO/UNESCO soil map. He concluded that the limits on the images which correspond fairly well to the limits of major soil units on the maps have been traced by using the gray tonality. The author also suggested potential applications for the Meteosat and GOES imageries based on broad synoptic character and the high repetition. These characteristics are important for studying the dust transports and deposits, moving dunes and forest advances or retreats [17].

Gombeer [16] stated that using Meteosat data and reports from synoptic weather and rainfall stations, algorithms were developed to map rainfall, net radiation, evapotranspiration, soil moisture availability, thermal inertia, and germinations. These algorithms were applied to a test area in Mali to map and monitor these variables during a test period of 18 days. The results demonstrated the ability of Meteosat to provide information in areas where round stations are scares or existing measurements unreliable.

Practically, Meteosat was used to study a thunderstorm case over the eastern desert cliffs, Qena area, Egypt on April 13 and 14, 1985. The thunderstorm occurred as a result of occasional monsoon low movement from the Sudan to the Gulf through Egypt as following sequence:

1. On April 11, 1985, instability of weather started in Qena area, related to the advancing of both the cold front coming from west and the Sudanese monsoon low from South (**Figure 9**).

2. On April 12, 1985, the Sudanese monsoon low continued to advance northwards and reached latitude 20, while the entire area was dry (**Figure 10**).

3. On April 13, 1985, the Sudanese monsoon low staked to a heavy cloud over the Qena area, which caused the thunderstorms. The dancing of the cold front from the west has helped in pushing the cloud toward the sea mountain and its ascending resulted in the flash rain (**Figure 11**)

This thunderstorm has caused the destruction of Khozam dam, which was built to protect the village of Khozam against the thunderstorm. It resulted also in lot of extreme damage in the village where the water reached a height of 2.25 m.

5. Evaluation of soil degradation in Northern Sinai (Egypt), using remote sensing and GIS

Soil degradation, as defined by FAO, UNEP [18] is "a process which lowers the current and/or the potential capability of soil to produce (quantitatively and/or qualitatively) goods or services." In 1975, UNEP, FAO, and UNESCO develop a methodology for assessing soil degradation on a global scale. The methodology was tested in America north of the equator and in the near and Middle East.

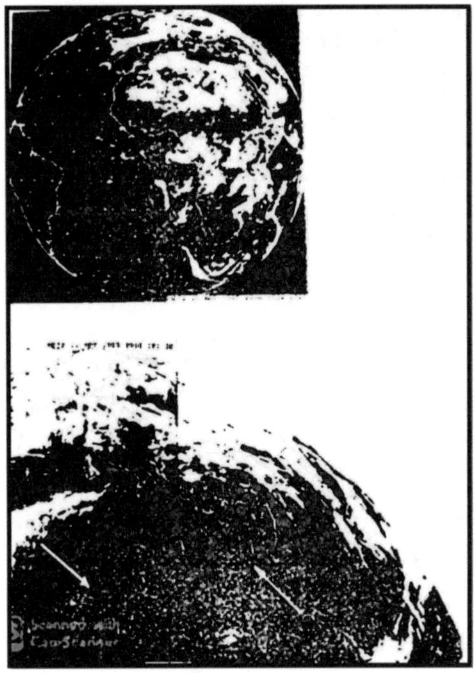

Figure 9.
Meteosat of April 11, 1985. It shows the advancing of the Sudanese Monsoon from the South toward the Red Sea. At the same time, the cold front started to move from the west the east.

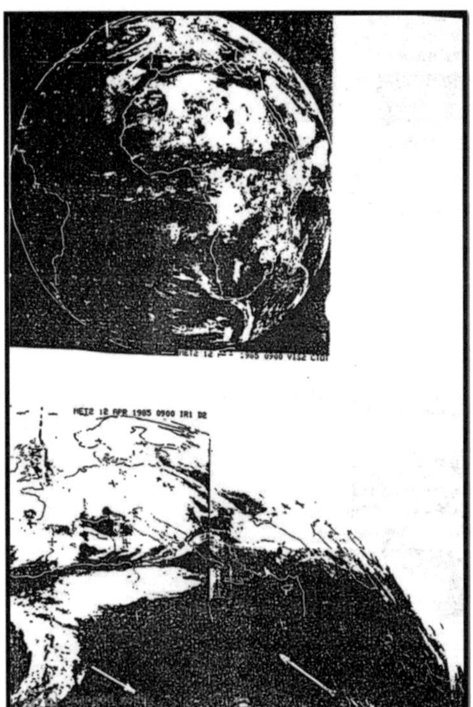

Figure 10.
Meteosat of April 12, 1985. It shows the continuous movement of the Sudanese Monsoon to the latitude of 20N, while the entire area was dry. The same time the cold front started to move from the west the east. Also continue moving eastwards.

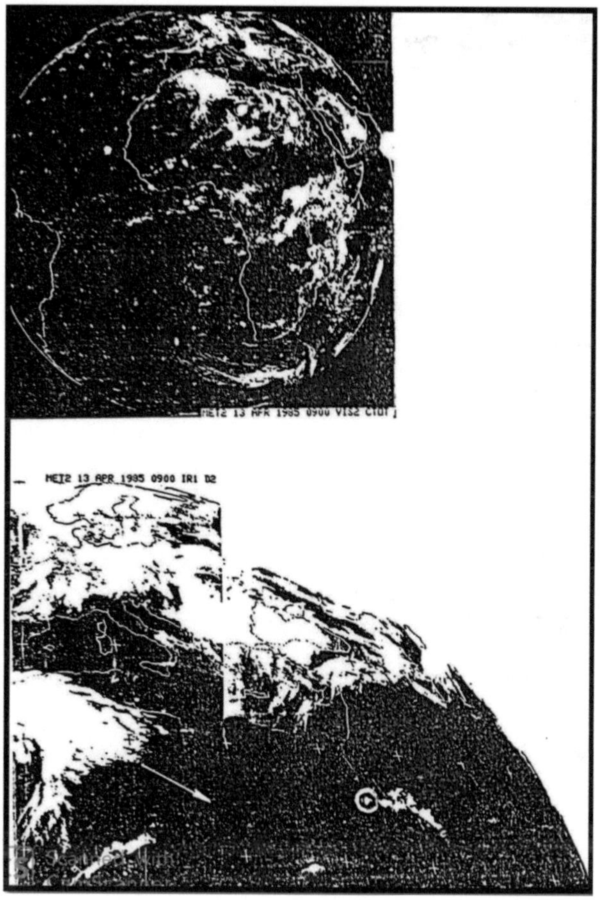

Figure 11.
Meteosat of April 13, 1985. The sticking of the Sudanese monsoon with clouds that caused the thunderstorms.

The recorded results are illustrated on four maps published at a 1:5,000,000 scale. It was clearly confirmed that soil degradation occurs over extended areas, which are not fully degraded then, but threatened by expected future degradation. Duly aliened by these results, FAO and UNEP [19] provided initiative to refine the original methodology to better serve scientists and managers in assessment of damage already done and future threats to land. The provisional methodology, published in 1983 has been developed to be scale independent, so that it may be applied at a global, national regional/provincial, and local project planning levels. It is designed to provide map able data that may potentially be used to plan strategies to conserve the remaining productive soil and to prevent soil degradation in areas not then affected.

The current study aims to assess soil degradation in north Sinai region, applying the above mentioned provisional methodology at 1:250,000 mapping scale. This scale is considered appropriate for planning at national level.

The Sinai Peninsula covers an area of 61,000 km^2, representing around **6%** of Egypt's territories. It represents a promising and strategic region for economic development. Northern Sinai region has considerable potential for agriculture, fisheries, and summer resorts. Much of the arable land in this area would eventually be irrigated with Nile river water through the El-Salam canal [20]. The objectives of the study are to assess existing and potential risk of soil degradation. The following research aspect also highlights a remote sensing and GIS practitioner's viewpoint:

- To test soil degradation assessment methodology at 1:*250,000* scale
- To use digital image processing and GIS techniques to derive input to the assessment model
- To evaluate the procedures and results

5.1 FAO concept of soil degradation monitoring and evaluation

Land degradation processes are phenomena that result in soil quality diminutions, leading to a risk of lowering current or potential productivities [21]. The present state of soil degradation is derived from the risk values by introducing the human activity represented by land use and soil management. Although often interacting, the soil degradation processes may be grouped into six categories, which are: water erosion, wind erosion, excess of salts, chemical degradation, physical degradation, and biological degradation. The current study assesses four of the six degradation processes, excluding only chemical and biological degradation. Soil degradation is expressed in the "FAO/UNRP and UNESCO provisional methodology" in units appropriate for each process. For example, soil erosion by water or wind is expressed by "soil loss in ton/ha/year" and salinization by "increase of EC in dS/m" and physical degradation by "increase of bulk density in g/cm^3/year." The degradation hazard values are then classified compared to listing of different soil degradation classes [22].

5.2 Assessment methodology and data sources

A parametric formula, based on the universal soil loss equation (USLE), was used [23]. Input data values are derived from a combination of direct measures and information of remote sensing and thematic maps. The formula can be expressed in the general form as:

$$D = .f(C, S, T, V, L, M)$$

where D is degradation, C is climatic aggressively factor, S is the soil factor, T is the topographic factor, V is the natural vegetation factor, $\boldsymbol{L}$ is the land use factor, and M is the management factor. For each degradation process, a similar formula is used. The values of the variables are chosen in such a way that solving of the equation gives a numeric indication of the degradation rate. The formula describes the processes only approximately and the values assigned to each factor are approximate in the present state of knowledge. Thus, the final results should not be regarded as absolute values for soil loss but as indication of the magnitude of degradation [24].

ArcGIS software is used to manage and manipulate the thematic map data, processed satellite images, and tabular data sources. ERDAS IMAGIN digital image processing programs are utilized to process the images, including radiometric and geometric correction, and to derive values of Normalized Difference of Vegetation Index (NDVI).

IDRISI software is employed with ***ArcGIS*** to generate slope values [25, 26].

5.3 Special processing concerns

Most of the data required the parametric formula were derived from map data sources or published data. The management factor or M value is derived from digitally processed satellite data.

The NDVI values are computed for each pixel in two 1024 × 104 datasets from 1984 Landsat MSS (**Figure 12**) and 1990 Landsat TM imagery (**Figure 13**). Iso-clustering classification is used to classify the NDVI images into six vegetation density classes. The classified images then converted to ArcGIS format and crossed with the soil coverage. The composed dataset is thus used in establishing a vegetation density rating for each soil polygon, thus deriving a management factor.

The other problem involves generating the percent slope data which are used in developing the rating of topography or (T factor) in the USLE. The following multi-step process is followed:

1. First, the digitizing of contour lines from 1:100,000 scale plan metric maps (**Figure 13**). *ArcGIS line* coverage is developed from digitized data and attributed with appropriate elevations.

2. Second, contour data are clipped to the digitized soil coverage and exported to IDRISI. A copy of the soil coverage is imported to IDRISI.

3. Third is the development of a digital elevation model (DEM) in IDRISI vector contour data.

Figure 12.
Processed Landsat MSS—of Wadi El-Arish region, Sinai case study.

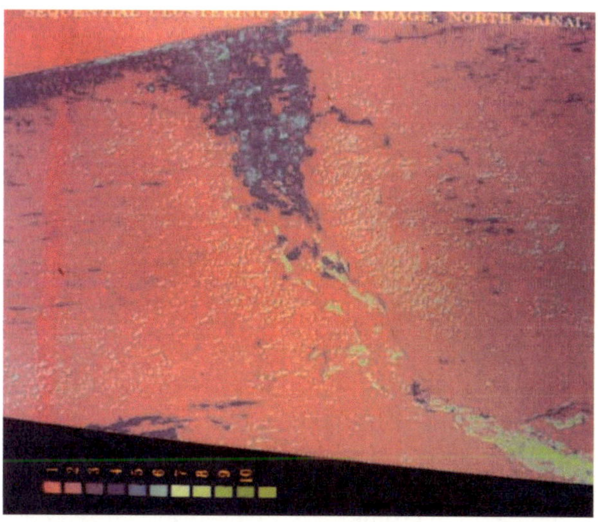

Figure 13.
Processed Landsat ETM—of Wadi El-Arish region, Sinai case study.

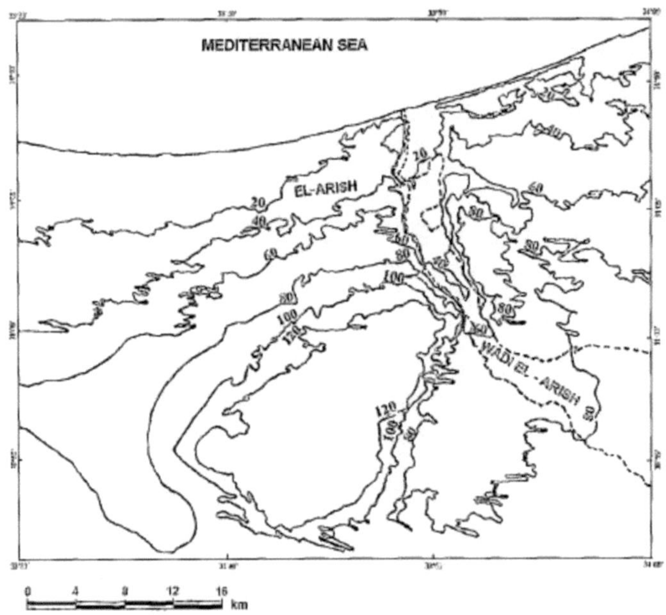

Figure 14.
Contour map of the Arish, Sinai study area.

4. Fourth, a percent slope value is generated for each cell.

5. Fifth step is to link a composite image of the percent slope data and the soil data.

6. Sixth and final step use the histogram to determine the distribution of slope values for each soil unit and to assign a percent slope class to each soil polygon.

5.4 Generation of the model

The soil coverage [20, 27] developed in ArcGIS is the base for generating the model (**Figure 14**). Climatic ratings, soil factors, topographic, vegetation, land use, and management ratings are added as attributes to the polygons of the soil coverage. Four soil profiles, representing different soil types in the study area, were investigated. Soil samples were collected and laboratory analyzed for calculating the soil factor acc. USLE. Computing the parametric equations is then completed for the four soil degradation processes, both for current soil degradation and the risk of soil degradation.

6. Results and discussions

Tables 1 and **2** show the laboratory analyses of the collected soil samples. Determination of the soil factor in the USLE is based on the results of these analyses. The soil erodibility factor for water erosion is calculated from Wischmeier's nomograph [28]. A correlation between soil texture and the wind erodibility was used [29]. Soil texture and depth to ground water have been utilized for rating the soil factor in salinization. The silt clay ratio is considered as an important factor contributing to the physical degradation process, as shown in **Figure 15** [24].

Profile no.	Depth (cm)	Mechanical analysis (mm)						Texture	O.M.	Structure
		2–>1	1–>0.5	0.5–>0.25	0.25–>0.125	0.125–>0.63	<0.63			
1	0–30	0	2	46.5	47.0	2	2.5	Sand	0.1	Loose
	30–60	0	1	50.0	46	2	1	Sand	nil	Loose
	60–100	0	5.5	58.0	33.5	2	1	Sand	nil	Loose
	100–150	0	10.5	58.5	25.5	1	1	Sand	Nil	Loose
2	0–40	0	1	11	82	3.5	2.5	Sand	0.1	Loose
	40–100	0	2.5	26.5	69	1	1	Sand	Nil	Loose
	100–150	0	7.5	18	72	1.5	1	Sand	Nil	Loose
3	0–35	0	3	10.1	35.1	22	9.8	S. loam	0.5	w. subang.
	35–70	0	1	2.90	50	27	17.5	S. loam	0.4	w. subang
	70–120	0	3.5	5.00	39.6	31.2	20.7	Loam	0.2	mod subang
4	0–60	0	2	39	48	7.5	5.5	Sand	0.4	Loose
	65–105	0	3	29.2	21.6	13.3	31.9	S. clay	−0.5	m.m. subang
	105–140	0	3.6	32.0	25.9	14.2	19.3	S. loam	0.3.	w.m. subang
	140–175	0	4.7	33.0	52.2	0	10.1	Sand	0.2	Loose

Table 1.
Some physical soil properties of selected soil profiles.

Profile no.	Depth	EC (mmohs/cm)	$CaCO_3$ (%)	Soluble salts (mequiv./1)							
				Cations				Anions			
				Ca	Mg	Na	K	CO_3	HCO_3	Cl	SO_4
1	0–30	0.5	6.3	1.7	0.9	2	0.5	0	0.8	2.8	1.5
	30–60	0.4	8.0	1.4	0.5	1.9	0.5	0	0.7	2.4	1.1
	60–100	0.5	10.1	2.1	0.3	1.9	0.4	0	0.7	2.5	1.5
	100–150	0.6	10.2	1.1	0.6	3.5	0.3	0	0.8	3.1	1.6
2	0–40	0.7	2	0.7	0.5	5.8	0.1	0	0.8	4.5	1.8
	40–100	0.8	0.3	0.6	0.4	7.5	0.2	0.5	0.9	6.1	1.2
	100–150	0.9	0.3	0.8	0.3	7.7	0.2	0.5	0.9	6.2	1.4
3	0–35	0.7	43.5	1.6	0.6	4.6	0.2	0	1.2	4.2	1.8
	35–70	1.1	54.5	1.8	1	7.9	0.2	0	1.3	5.8	3.8
	70–120	2.2	52.5	3.4	2.5	17	0.2	0	2.9	16.2	3.9
4	0–65	0.6	5.5	1.9	0.5	3.3	0.4	0	0.9	3.6	1.6
	65–105	2.5	28.5	2.6	2	21	0.2	0	3.1	18.4	4.4
	105–140	8.3	29.5	17	11.8	69	0.5	0	7.4	72	19.3
	140–175	6.1	23	12	8.9	50	0.4	0	7.4	43.8	19.8

Table 2.
Some soil chemical properties of selected soil profiles.

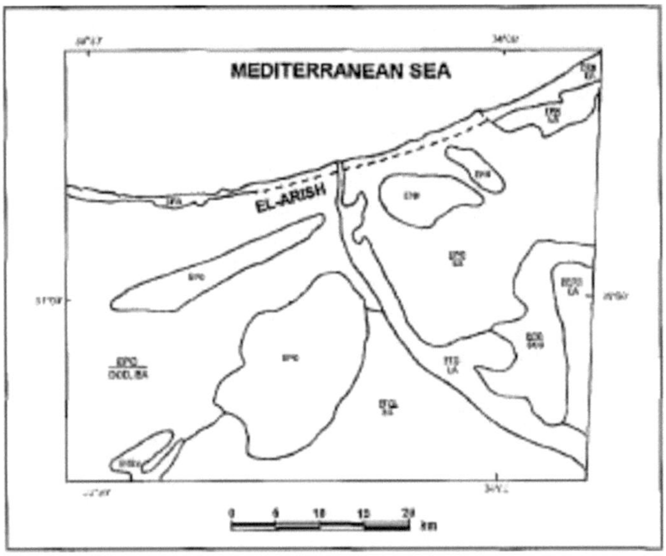

Figure 15.
Soil map of the Wadi El-Arish region, Sinai study area.

7. Water erosion

Table 3 shows the values of risk and present status of water erosion and the input parameters for their calculation. The area is generally exposed to a non to slight risk as the sand fraction is dominant in most soil types (**Table 1**). However,

Area (km^2)	ID	Soil type	Climatic factor	Soil factor	Topo. factor	Human factor	Risk	Present state
000051.130	1	EPA	16.39	0.03	0.35	0.32	0.1721	0.0551
000062.810	2	EPQ	16.39	0.03	2.00	0.45	0.9834	0.4425
000212.500	3	EPQ	16.39	0.03	2.00	0.45	0.9834	0.4425
000582.190	4	EPQ	16.39	0.03	2.00	0.45	0.9834	0.4425
000017.190	5	EHSS	16.39	0.08	0.35	0.07	0.4589	0.032 1
000301.875	6	EFCL/EA	16.39	0.08	0.35	0.07	0.4589	0.0321
000126.560	7	EFD/EA	16.39	0.08	0.35	0.07	0.4589	0.0321
000041.880	8	EPN	16.39	0.14	3.50	0.12	8.03 1 1	0.9637
000014.380	9	EPN	16.39	0.14	3.50	0.12	8.031 1	0.9637
000030.310	10	EPN/EA	16.39	0.14	3.50	0.12	8.0311	0.9637
000022.190	11	EPN/EA	16.39	0.14	3.50	0.12	8.03 11	0.9637
624514.990	12	EPQ/EA	16.39	0.03	2.00	0.45	0.9834	0.4425
000109.380	13	EPQ/ DOD	16.39	0.03	2.00	0.45	0.9835	0.4425
000053.750	14	EFGS/EA	16.39	0.08	0.35	0.07	0.4589	0.0321

Table 3.
Values of risk and present status of water erosion and the input parameters for their computation.

values of water erosion are the highest in the *Normipsamments* soils (EPN) as the topsoil is characterized by a sandy loam texture (soil profile no. 3) as opposed to the other soil units with sandy top soils.

8. Wind erosion

Table 4 shows the values of risk and present status of wind erosion and the input parameters for their calculation. Annual average of wind velocity, in El-Arish station reaches 4.30 Knots (2.214 m s^{-1}). Thus, the wind erosivity factor is high (50–150) in the study area. All soil types are characterized by high (50–200 t h^{-1} $year^{-1}$) to very high (>200 t h^{-1} $year^{-1}$) risk values of soil loss by wind erosion. Wind erosion in the study area is particularly important because the soils are mostly dry and the vegetation cover is scattered or absent. Cultivation of barley and non-conventional crops in some soils reduces the present state of wind erosion hazard. However, wind erosion is more pronounced in the *Psamments* soils (Quartzipsaments EPQ and *Aquipsamments* EPA), which are formed on sand dunes.

9. Salinization

Table 5 shows the values of risk and presents status of salinization and the input parameters for their calculation. The study area is characterized by hyper arid climatic conditions: the precipitation (P) is less than 1/3 of the potential evapotranspiration (PET) and at least one 12 month period without rainfall. Thus, the proposed climatic index (PET/P) is very high (0.5–3.3).

The present state and risk values in the *Psamments* (i.e., EPA, EPQ, and EPN) are slight to moderate. The coarse texture and rapid permeability of these soils reduce

Area (km^2)	ID	Soil type	Climatic factor	Soil factor	Topo. factor	Human factor	Risk	Present state
000051.130	1	EPA	100	3.5	1	0.70	350	245.00
000062.810	2	EPQ	100	3.5	1	1.00	350	350.00
000212.500	3	EPQ	100	3.5	1	1.00	350	350.00
000582.190	4	EPQ	100	3.5	1	1.00	350	026.25
000017.190	5	EHSS	100	1.75	1	0.15	175	026.25
000301.875	6	EFCL/EA	100	1.75	1	0.15	175	026.25
000126.560	7	EFD/EA	100	1.75	1	0.15	175	052.50
000041.880	8	EPN	100	1.75	1	0.30	175	052.50
000014.380	9	EPN	100	1.75	1	0.30	175	052.50
000030.310	10	EPN/EA	100	1.75	1	0.30	175	052.50
000022.190	11	EPN/EA	100	1.75	1	0.30	175	052.50
624514.990	12	EPQ/EA	100	3.5	1	1.00	350	350.00
000109.380	13	EPQ/ DOD	100	3.5	1	1.00	350	350.00
000053.750	14	EFGS/EA	100	1.75	1	0.15	175	026.25

Table 4.
Values of wind erosion present status and risk and the input parameters for their computation.

Area (km^2)	ID	Soil type	Climatic factor	Soil factor	Topo. factor	Human factor	Risk	Present state
000051.130	1	EPA	1.50	0.1	1.0	0.7	0.150	0.105
000062.810	2	EPQ	1.50	0.1	1.0	0.5	0.150	0.075
000212.500	3	EPQ	1.50	0.1	1.0	0.5	0.150	0.075
000582.190	4	EPQ	1.50	0.1	1.0	0.5	0.150	0.075
000017.190	5	EHSS	1.50	1.0	5.0	0.7	7.500	5.250
000301.875	6	EFCL/EA	1.50	1.0	5.0	0.7	7.500	5.250
000126.560	7	EFD/EA	1.50	1.0	5.0	0.7	7.500	5.250
000041.880	8	EPN	1.50	1.0	0.1	0.7	0.150	0.105
000014.380	9	EPN	1.50	1.0	0.1	0.7	0.150	0.105
000030.310	10	EPN/EA	1.50	1.0	0.1	0.7	0.150	0.105
000022.190	11	EPN/EA	1.50	1.0	0.1	0.7	0.150	0.105
624514.990	12	EPQ/EA	1.50	0.1	1.0	0.5	0.150	0.075
000109.380	13	EPQ/ DOD	1.50	0.1	1.0	0.5	0.150	0.075
000053.750	14	EFGS/EA	1.50	1.0	5.0	0.7	7.500	5.250

Table 5.
Values of risk and present status of salinization and the input parameters for their computation.

the risk of salinization. *Haplorthents* (EHSS) and *Fluvents* (EFGS, EFD, and EFCL) are exposed to a very high risk of salinization. This is related to their shallow soil profiles, fine texture, and medium to low permeability. Since *Fluvents* occur in wadis, valley floors, desert basins, and playas, the topographic factor increases the salinization risk. Furthermore, agricultural practices on these soils, especially the excessive application of irrigation water, increase salinization hazard.

10. Conclusion and recommendation

Management and planning agricultural expansion in desert areas are essential for self-sufficiency of food production. However, many degradation processes, which are severe environmental extremes, are active on the soil and cause deterioration in their potential productivity. The evaluation and control of soil degradation and productivity are based on environmental information. GIS techniques are useful in storing, retrieving, and manipulating such information. Furthermore, remote sensing techniques GIS are useful in updating the status of soil deterioration and providing services to a risk model. It should be advised that the final values generated by parametric equations are not absolute values of soil loss. These values merely give an approximate indication of the likely magnitude of degradation. Additionally, better sources of data for the management factor and percent slope information need to be identified.

Author details

Abd-alla Gad
National Authority of Remote Sensing and Space Sciences (NARSS), Egypt

*Address all correspondence to: abdallagad1@gmail.com

IntechOpen

© 2020 The Author(s). Licensee IntechOpen. Distributed under the terms of the Creative Commons Attribution - NonCommercial 4.0 License (https://creativecommons.org/licenses/by-nc/4.0/), which permits use, distribution and reproduction for non-commercial purposes, provided the original is properly cited. BY-NC

References

[1] Gómez F. Extreme environment. In: Gargaud M et al., editors. Encyclopedia of Astrobiology. Berlin, Heidelberg: Springer; 2011

[2] Taminiau J, Nyangon J, Lewis AS, Byrne J. Sustainable business model innovation: Using polycentric and creative climate change governance. In: Fields Z, editor. Collective Creativity for Responsible and Sustainable Business Practice. IGI Global; 2017. pp. 140-159

[3] UNCCD. Article 2 of the Text of the United Nations Convention to Combat Desertification. 1994. Available from: http://www.unccd.int/Lists/Site DocumentLibrary/conventionText/ conv-eng.pdf

[4] Westin FC. Soil and land resource inventory using Landsat data. In: Remote Sensing for Resource Management. Ankeny, Iowa: Soil Conservation Society of America; 1982. pp. 243-254

[5] Hill E. A deep crustal shear zone exposed in western Fiordland, New Zealand. Tectonics. 1995;**14**:1172-1181

[6] FAO/UNEP and UNESCO. A Provisional Methodology for Soil Degradation Assessment. Rome: FAO; 1999

[7] Miller JD, Hutchins M. The impacts of urbanisation and climate change on urban flooding and urban water quality: A review of the evidence concerning the United Kingdom. Journal of Hydrology: Regional Studies. 2017;**12**:345-362. DOI: 10.1016/j.ejrh.2017.06.006

[8] Binns T. Is desertification a myth? Geography. 1990:106-113

[9] Helldén U. Desertification: Time for an assessment? Ambio. 1991;**20**(8): 372-383

[10] Hill J, Hostert P, Tsiourlis G, Kasapidis P, Udelhoven TH, Diemer C. Monitoring 20 years of increased grazing impact on the Greek island of Crete with earth observation satellites. Journal of Arid Environment. 1998;**39**: 165-178

[11] Thomas DSG, Middleton NJ. Desertification: Exploding the Myth. Chichester: Wiley; 1994

[12] Barrett EC, Hamilton MG. The use of geostationary satellite data in environmental science. Progress in Physical Geography. 1982;**6**(2):159-214

[13] Darnll WI, Harris RC. Satellite surface capability for surface temperature and meteorological parameters over the oceans. International Journal of Remote Sensing. 1983:65-92. 1pp

[14] Eigenwillig N, Fischer H. Determination of midtropospheric wind vectors by tracking pure water vapor structures in METEOSAT water vapor image sequences. Bulletin of the American Meteorological Society. 1982; **63**:44-58

[15] Robson A, Morgan J, Herschy RW, Zchav J. Detection of natural disasaters via Meteosat. ESA Bulletin. 1982;**29**: 10-18

[16] Gombeer R. Crop and land use classification study on SAR and MSS data over Belgium ESA SAR 580 Invest. Preliminary Report. 1983

[17] Nicholson S, Jinnah S, Gillespie A. Solar radiation management: A proposal for immediate polycentric governance. Climate Policy. 2018;**18**(3):322-334. DOI: 10.1080/14693062.2017.1400944

[18] FAO/UNEP. Guidelines for the Control of Soil Degradation. Rome, Italy: FAO; 1983

[19] FAO/UNEP. Report on the FAO \UJNEP expert consultation on methodology for assessing soil degradation. In: Project no. 1106-75-05. Rome, Italy: FAO; 1978

[20] Hammmi M. Land master plan region report—Sinai, Cairo, Ministq of Development, General Authority for Rehabilitation Project and Agricultural Development. 1986

[21] Lal R, Stewart A. Soil Degradation. New York: Springer-Verlag; 1990

[22] FAO/JNEP, UNESCO. A Provisional Methodology for Soil Degradation Assessment. Rome, Italy: FAO; 1979

[23] Soil Science Society of America (SSSA). Universal soil loss equation: Past, present and future. In: SSSA Special Publication Number 8. Madison, Wisconsin: SSSA; 1979

[24] Shalaby A, Ali RR, Gad A. AGIS-based model for desertification sensitivity assessment, case study: Inland Sinai and Eastern Desert Wadies. International Journal of Basic and Applied Sciences. 2011;**1**(1):1-11

[25] Gad A. Study of soil degradation processes in the eastern Nile Delta using GIS and Remote sensing. Egyptian Journal of Remote Sensing. 1993;**1**

[26] Gad A, Younes, Abdel-Hadmy A. Assessment of soil degradation processes in the middle part of the Nile Valley, Egypt, using GIS and remote sensing techniques. In: Proceedings of the 15th World Congress of Soil Science, Acapulco, Mexico, 10-16 July. 1994. pp. 217-218

[27] Abdel-Samae G, Labibt M, Abdel-Hady MA. Soil classification and potentials in Sinai peninsula from Landsat images. In: Proceedings of the International Symposium on Remote Sensing in Arid and Semi-arid Lands; Cairo, Egypt, January. 1982

[28] Kim JH, Keane TD, Bernard EA. Fragmented local governance and water resource management outcomes. Journal of Environmental Management. 2015;**150**:378-386

[29] Klmberlm LW, Hidlebaugha L, Grimewalad R. The potential wind erosion problem in the United States. Transactions of ASAE. 1977

Chapter 11

Scaling Up Sustainable Biofuels for a Low-Carbon Future

Tahira Shafique and Javeria Shafique

Abstract

Fossil fuels oil, coal, and gas are valuable resources that are depleting day by day around the world and also imparting a negative impact on the environment. Biofuel because of its dynamic properties; its market values; and being sustainable, renewable, biodegradable, economic, non-pollutant, and abundant is an alternate source of energy. Each country can produce it independently, and because of these valuable properties biofuels have become superior over fossil fuels. This chapter gives a concise preface to biofuels and its impact on the environment. It includes definitions; classifications; impact on environment; implications; types of production techniques like chemical, biochemical, physical, and thermochemical techniques; types of resources like lignocellulosic-biomass, feedstock energy crops, algae, micro-algae, all kinds of solid wastes; and biofuels of prime importance like solid biofuels (biochar, solid biomass), gaseous biofuels (biogas, bio-syngas, and bio-hydrogen), and the most important liquid biofuels (bioethanol, biodiesel, and bio-oil). Due to increasing global warming and climate-changing conditions, in the near future biofuel being an environment-friendly resource of energy will be a substantial part of the world's energy demand, with no or zero polluting agents.

Keywords: biofuels, biodiesel, renewable energy resources, global warming solutions, alternate of fossil fuels, energy investment, future fuels, biodiesel, ethanol, oils, TAGs, biogas, green-diesel, algae, macroalgae, plants, wood

1. Introduction

Biofuels: a term that has broad implications in regard to the world's needs, biofuels have been a part of human life for a long time ever since the first person burned wood to get heat. Over time, he found the value of heat and started cooking by burning wood. So, the biofuels have an extended history than humanity's record and civilizations because these were present in the form of grass and wood before when we human beings found and used the fire.

With the progression of human civilizations, biofuels took an important place in the human's life. For example, CH_4 gas also known as domestic gas or biogas was used to warm water to take a bath in the 10th BC [1]; CH_3CH_2OH or bioethanol was used in a combustion engine in 1826 [2]; Sadi's Carnot engines used biodiesel also known as Rudolph's diesel [3]; and for the very first time, peanut oil was used in the busses and trucks in 1896 [4]. Many other such applications developed and changed people's lives around the world [5].

In contrast with fossil fuels natural gas, coal, and oil, biofuels have some advantages:

IntechOpen

- These are renewable [6]
- Resources of biofuels are abundant [6]
- Biofuels do not play any role in ozone layer depletion as these are neutral to greenhouse gases [7]
- Sulphur oxide emissions are zero or negligible [8]
- They result in fewer nitrous oxide emissions [9]
- Biofuels are friendly to the environment [10]
- These are easy to produce [11]
- All biofuels are biodegradable [12]
- All kinds of biofuels are sustainable [13]
- Biofuels production procedures are safe [14]

Biofuels also have some economic benefits:

- These fuels will lead to rural and agricultural development [15]
- These fuels enhance supply range [16]
- They lead to reduced dependence of energy on imports [14, 17]
- These fuels increase opportunity of jobs [14, 18]
- They improve rural economy [16].

To increase and promote the biofuel industry, several countries have planned many rules and objectives. Here, documents issued in the European Union, USA, Brazil, Thailand, India, and China are mentioned in **Table 1**.

Country	Target or mandate	References
EU	Mandate: minimum of 10% of transport fuel from renewable fuels by 2020	[19]
United States	Mandate: 36 billion gallons of biofuel by 2022	[19]
Brazil	Mandate: biodiesel use set at 10% by 2020	[19]
India	Indicative: 20% blending for both ethanol and biodiesel by 2017	[19]
China	Target: solid biofuels (10,000 t/year), biogas (billion m^3/year), nonfood bioethanol (10,000 t/year), biodiesel (10,000 t/year) by 2020	[19]
China	Target: 12.7 Bnl ethanol and 2.3 Bnl biodiesel consumption in 2020 15% of fuel consumption to be non-fossil fuel by 2020	[19]
Thailand	Ethanol: E20 mandatory since 2008 Biodiesel: B2 mandatory since 2008 and B5 since 2012	[19]

Table 1.
Set of policies ordered by global biofuel producers.

2. Biofuel

2.1 Definitions

The term fossil fuels was used for fuels like natural gas, coal, and oil gotten from the decomposition of different plants and a variety of animals that remain buried underneath the earth's deepest layer for many years [14]. Over time, the word "biofuel" started being used. Till now, there are many distinctive definitions for the word "biofuel." Some of these statements are

- Bio-fuels are obtained from or with the help of microorganisms [14].
- Biofuels are those fuels that are obtained from biomass, petroleum, natural gas, and firewood [20].
- We derive these from different biological and plant materials [21].
- Bio-fuels are the products obtainable from natural resources, which include timber, peat or bagasse, or transformed chemically from biomass to synthesize ethanol, charcoal, bio-oil, and biogas [4, 22].
- These fuels are extractable from biomass feedstock or waste materials [15].
- Bio-fuels include liquid, solid, and gaseous fuels obtained from organic material like plants and animals [19].
- These are also known as the renewable fuels like biodiesel, bio-hydrogen, charcoal, bioethanol, and biogas obtained from biomass gotten from the organic waste materials and are reliable for transportation [10, 19, 23].
- Liquid biofuels: these include biodiesel extracted from fats and oils, bioethanol extracted from lignocellulosic material, sugar and starch, etc. [18].

According to the above definitions, "biofuel" can be defined as any kind of fuel that is produced by or from any renewable living organisms.

2.2 Significance

"Biofuels" or "bio-fuels" are significant in many ways such as:

- Biofuel is employed to supply heat, energy, light, and power.
- Biofuel is also claimed as a living fuel, that is produced from different living micro/macro-organisms: (the "bio" springs from the Greek word "bios," means "life"); these are the ways in which these fuels differ from all other fuels like fossil fuels having hydrogen and carbon.
- Biofuels are renewable, therefore the living organism can reproduce it in a relatively short time. While fossil fuels took many years and are non-renewable (this property of biofuels made it significant over fossil fuel).
- Biofuels might be obtained from plant structure, water algae, micro-algae, manure, animal waste, sludge, etc.

- Biofuel could be a solid, gas, or liquid.
- One type of biofuel is interconvert able to another type of fuel.

2.3 Classification

As we know there are many definitions of "biofuels"; similarly there are many classifications of biofuels according to different eras and research groups. According to the very first classification based on commercialization position of the biofuel production and resources, these are usually classified into two groups: one is conventional biofuels and the other is advanced biofuels (**Figure 1**). Commercialization can be ranked accordingly: research < demonstration < early commercial < commercial. The conventional or traditional bio-fuels contain such biofuels that are based on biodiesel gotten from the trans-esterification process; bioethanol obtained from starch and bio-methane or biogas from animal waste obtained via anaerobic digestion process. With regard to advanced bio-fuels, an early commercial-stage biofuel is a hydrologically treated oil. The demonstration level and research level stages contain lignocellulosic or cellulosic bioethanol, BtL (biomass-to-liquid) biodiesel, bio-hydrogen, and biodiesel obtained from micro-algae. We know that advanced biodiesel is not dominant at commercial level but because of its environmental and economic demand, it will be dominate in the market and will be fully commercialized in more advanced form in near future. Based on feedstock resource availability and synthetic techniques, many researchers classify this biofuels into two groups like 1st-generation and 2nd-generation biofuels [24, 25]. This classification is shown in **Figure 2**. Biofuels of 1st generation contain biodiesel, biogas, and bioethanol [24, 26]. The unsaturated and saturated edible plant-oils obtained from the corn, soybean, sunflower, canola-seeds, and some oils containing fruits like palm, olives, and coconut are categorized under liquid bio-fuels [25, 26]. It also has been stated in different researches that the solid biofuels like agricultural waste material, dried manure, and fire-wood are also part of the 1st-generation biofuels. The lipid-derived bio-fuels obtained from the waste of vegetable-oils, insects, animal fats, and oil-producing microbes are categorized under biofuels of 2nd generation [25]. At present biofuels of 1st generation are fully commercialized, while 2nd-generation biofuels are still under developmental stages [24].

Biofuel's generations according to many other studies are classified in 1st, 2nd, and 3rd-generation biofuels [10, 27] mentioned in **Figure 3**. Biodiesel, biogas, bio-alcohols, bio-syngas, and vegetable oil, which are commonly known as biofuels

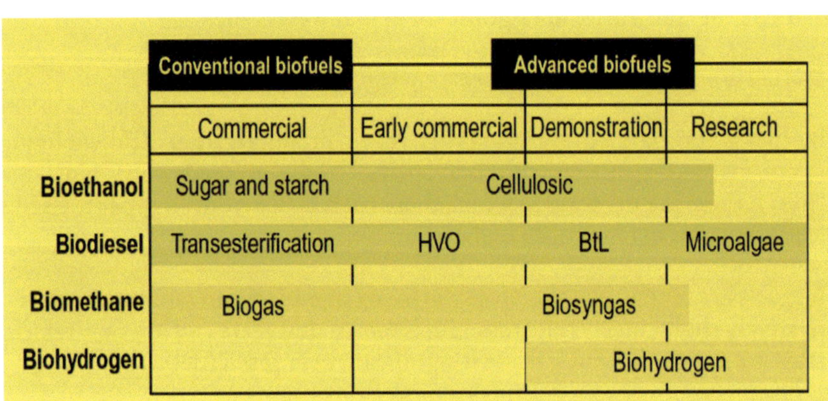

Figure 1.
Classification of biofuels and commercial status.

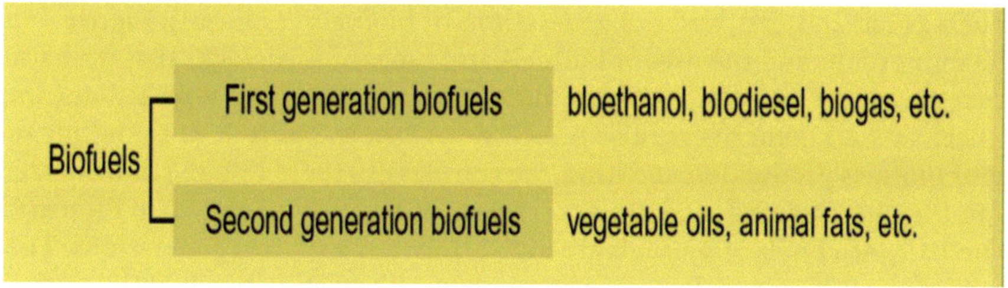

Figure 2.
Two generations of biofuels.

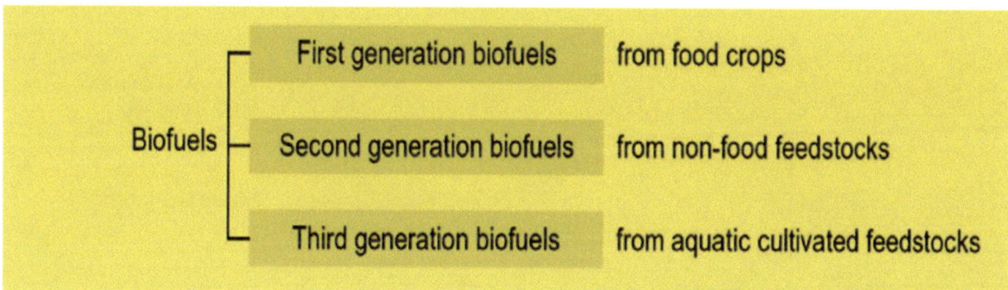

Figure 3.
Three generations of biofuel.

of 1st generation and obtained from edible crops: wheat, soybeans, vegetable oil, starch, corn, sugar, and grains, [28], are also contained within ryegrass, straw, wood, domestic refuse, switchgrass, grass cake, charcoal, grass cuttings, and dried manure [29]. The biofuels of 2nd generation includes biodiesel-FT(Fischer Tropsch), dimethyl ether, bioethanol, bio-syngas, biodiesel BTL(biomass-liquid) are synthesized from feedstock (lignocellulosic-biomass), nonfood material like cereal straw, wood grass, forest residues, sugar-cane, bagasse, energy crops (vegetative plants and short lifecycle forests) and domestics waste material [10, 30, 31].

The biofuels derived from feedstocks that are aquatically cultivated like cyanobacteria, algae, and micro-algae are 3rd-generation biofuels; these include biodiesel, bio-methane, bio-butanol, aviation fuels, bioethanol, jet fuels, gasoline, and vegetable oil [10, 32].

In another classification, biofuels are categorized into two groups: (i) primary biofuels included many solid biofuels without modifications like wood, grass, and wood slices that are burned for heating purposes and to cook different foods directly and (ii) secondary type biofuels have three generations of biofuels shown in **Figure 4** [33, 34].

According to some studies, biofuels are categorized into four groups: (i) natural bio-fuels are derived from simple organic materials like firewood, plants, vegetables, landfill gas, and animal waste and are used for heating, cooking, brick kiln, and production of electricity; (ii) the biofuels of 1st generation are obtained from eating able feedstocks, mostly palm, wheat, corn, soybean, maize, rapeseed, sugarcane, sugar beet, and oil crops [34]; (iii) the lignolytic feedstocks contain jatropha, miscanthus, sterculia, ceiba foetida, switchgrass, pentandra, and poplar are included in the 2nd-generation biofuels; (iv) algae feedstocks are the main source of 2nd generation biofuels **Figure 5** shows its classification [34].

With the advancement of research, oil-containing (jatropha) crops and nonfood (cassava) crops were known as one and a half- or 1.5-generation bio-fuels. Then biofuels took a new place in the classification tree and were re-classified into four groups: (1G) 1st-generation, (1.5G) 1.5-generation, (2G) 2nd-generation, and (3G) 3rd-generation of biofuels as shown in **Figure 6**. Some researchers grouped these bio-fuels as 1st, 2nd, 3rd, and 4th generations of biofuels [3, 35–40]. **Figure 7** shows all the four groups and the uses of biofuels and types of feedstocks that were used to produce these four generations of bio-fuels are represented in **Table 2**. Some important merits versus demerits are also briefly described in **Table 3**. With the advancement of biofuels production and kind, we can classify these into the five different groups: 0th, 1st, 2nd, 3rd, and 4th generations of biofuels, as shown in **Figure 8**.

The 1st to 4th types of biofuels are like the above-described generations. The biofuels of the 0th-generation biofuels are naturally existing bio-fuels like raw feedstocks and can be used directly with no processing, special treatment, or modifications.

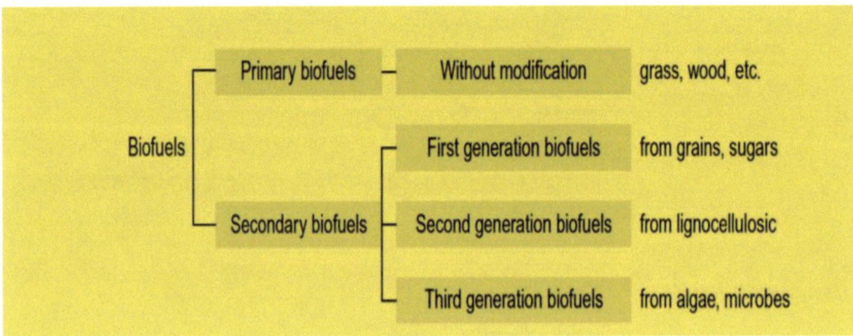

Figure 4.
Primary and secondary bio-fuels.

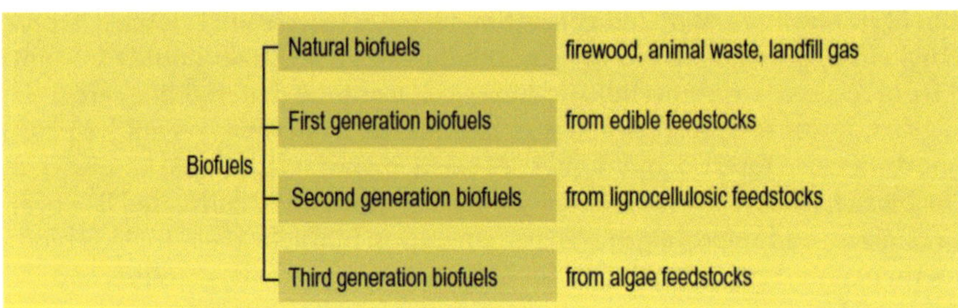

Figure 5.
Four groups of bio-fuels.

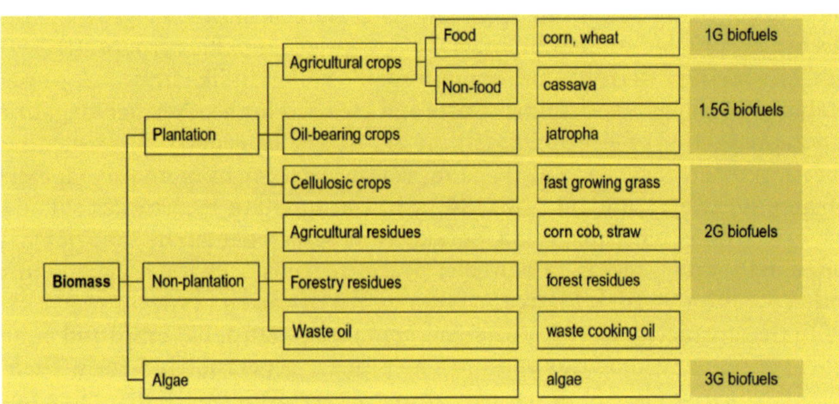

Figure 6.
Biofuels categorization into five generations.

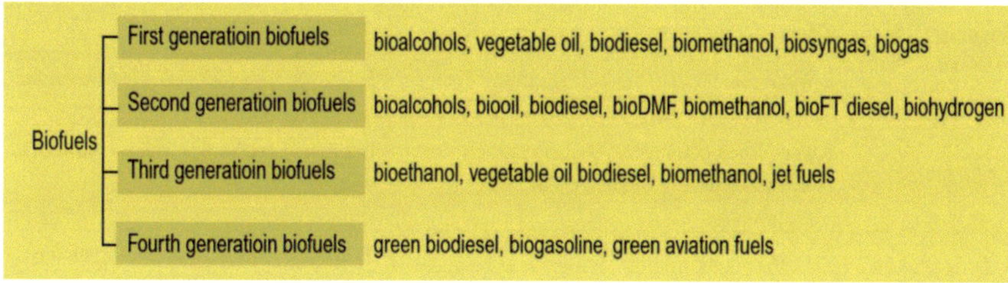

Figure 7.
Four groups classification of bio-fuels.

Biofuels	Examples	Feedstocks
1st generation	Bio-alcohols, vegetable oil, biodiesel, bio-methanol, bio-syngas, biogas	Sugar, starch, animal fats, soybean, soya, rapeseed, mustard, sunflower, maize, sugarcane, sugar beet, sorghum, potato, palm oil, coconut, canola, plant, cassava, castor, jatropha, sewage waste
2nd bio-alcohols, bio-oil, biodiesel, bio-DMF, generation	Bio-methanol, bio-Fischer-Tropsch diesel, bio-hydrogen	Nonfood crops, wheat straw, corn, wood, switchgrass, cereal straw, sugarcane bagasse, reed canary grass, forest residues, energy crops, municipal solid wastes, alfalfa, agave, jatropha
3rd generation	Bioethanol, vegetable oil, biodiesel, bio-methanol, jet fuels	Microbial species, algae, yeast, fungi, cyanobacteria
4th generation	Green diesel, bio-gasoline, green aviation fuel	Vegetable oil, biodiesel

Table 2.
Four generations of biofuels, feedstocks.

Biofuel	Merits	Demerits
1st generation	1. Biodegradable	1. Competition of land use
	2. Energy security	2. Blending with conventional fuel
	3. Feedstock can be easily produced by already existing infrastructure and technology	3. Highest carbon footprint compared with other generations of biofuel
	4. Environmental and social benefits	4. Requires large amount of inputs in terms of fertilizer, water and land area thereby reducing net energy ratio
	5. Feedstocks available at large quantities	5. Contributes to higher food prices owing to competition with food
		6. Might potentially have negative impacts on biodiversity
2nd generation	1. No competition with food	1. Even though requires less compared to 1st-generation biofuels, the land required for production of 2nd-generation feedstock is substantial
	2. Use of whole plant instead of only seeds or grains, and use of residues means more energy produced per hectare of land	2. The use of agriculture and forest residue degrades soil quality and also induces soil erosion
	3. Marginal lands can be used for planting of advanced feedstock such as *Jatropha* sp.	3. Complex processes are required
	4. Higher yield and lower land requirement	4. Low conversion as compared with petroleum fuel

Biofuel	Merits	Demerits
	5. Available feedstocks in large quantities	5. Conversion technologies are under development
	6. Feedstock can be easily produced by already existing infrastructure and technology	6. Lack of technological and research breakthrough
	7. Low cost for feedstock	7. Lack of efficient technologies for commercial applications
	8. Energy security	
	9. Production of high-value added products	
	10. Close to meeting the claimed environmental benefits	
3rd generation	1. No food or land competition	1. Difficult to harvest and process
	2. Produces more energy per acre than conventional crops	2. High processing cost
	3. Algae can be grown using land and water unsuitable for food production	3. Lack of technological and research breakthrough
	4. High oil yield	4. Not yet commercially feasible
	5. No toxic content	5. Production technology is under development
	6. Energy security	6. Requires new technologies from the production of feedstock to processing into final biofuel product
	7. Bioengineered algae are renewable	
	8. Low and sometimes no cost for feedstock	
	9. Improves performance of 1st- and 2nd-generation biofuels when employed in integrated biofuels	
4th generation	1. 4th-generation biofuel is argued to be carbon negative rather than simply carbon neutral, as it "locks" away more carbon than it produces	1. Lack of study on its practical performance in terms of technical and economic aspects
	2. Synthetic raw materials to produce biofuels is a possibility	2. High cost
	3. Energy security	3. Still in research and development stage
		4. Requires new technologies from the production of feedstock to processing into final biofuel product

Table 3.
Merits versus demerits of the all four generations of biofuels.

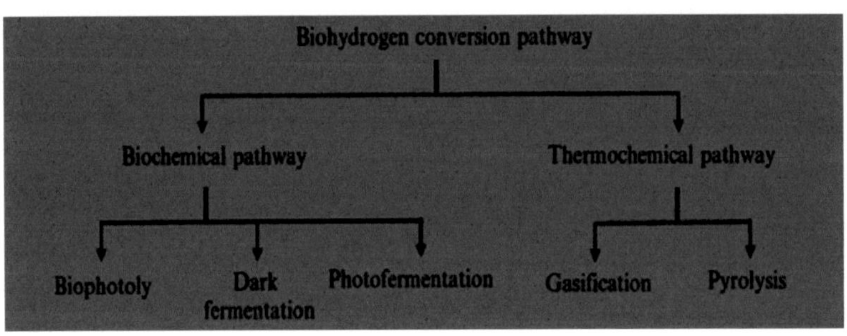

Figure 8.
Bio-hydrogen gas production pathways.

3. Gaseous bio-fuel synthesis techniques

Global warming and adverse climatic condition are propelling the researchers to bring a revolution with sustainable and renewable energy resources to reduce reliance on fossil fuels, which are depleting constantly. In this scenario, gaseous biofuels are environment friendly, will make vital contribution and will be a substantial part of the world's energy demand. The most valuable gaseous biofuels are bio-gas and bio-hydrogen.

3.1 Biogas (bio-methane)

A renewable, most sustainable and versatile energy source is biogas [41]. Shortly, biogas will make a remarkable impact to meet the energy demand. Biogas or bio-methane production resources are easily available at low cost and some even at no-cost like animal waste, domestic waste, animal feedstock waste, and municipal-organic, industrial effluent, for example, fat-separator wastes, glycerin, and food processing effluents waste material and residues of cereal crops [42, 43]. Biogas is the best replacement for fossil fuels [44]. It is safe to use in vehicles, generators, and combustion engines just like natural gas or fossil fuel gas.

The percentage of methane content in biogas defines its flammability and energy capacity. Production of biogas is mainly accomplished from the digestion of anaerobic biomass. It has 50–75% CH_4 (methane), 25–50% CO_2 (carbon dioxide), sulfur traces, water, hydrogen sulfide, oxygen, hydrogen, and ammonia. Production of biogas in methane treatment-plant that upgrades methane has similar properties just like natural gas and contains about 95% methane. To generate heat and electricity in co-generation units, de-sulfurized and dried bio-gas is safe to use [44]. With recent research, algae, water algae, and micro-algae are of prime importance for the production of bio-gas with outstanding results [42].

Anaerobic production of biogas is mainly divided into four steps: (a) hydrolysis, (b) acidification (also called acido-genesis), (c) de-hydrogenation/aceto-genesis, and (d) methanation [41]. As different anaerobic bacteria are used for biogas production, anaerobic conditions, pH, nutrient, and temperature are provided according to the demand for biogas production.

3.2 Bio-hydrogen

Bio-hydrogen is just simple hydrogen produced from the biochemical or thermochemical conversion of biomass like feedstock, residues of crops, cereal crops, agricultural grasses residues, livestock waste, forest biomass, waste-oil, algae, and micro-algae biomass, and industrial effluents [45, 46].

Hydrogen is a low-molecular mass gas having a higher heating value (HHV) of combustion products. The important thing that makes it perfect for the environment is that during chemical combustion after releasing energy, only water (H_2O) is the product. It has no harmful effect on the environment and global warming condition. Greenhouse gases are not produced during bio-hydrogen combustion. Bio-hydrogen production technologies are still under development [47, 48]. There are two techniques for bio-hydrogen production: (a) thermochemical (b) biochemical as shown in **Figure 8** [49].

3.2.1 Thermochemical method

Thermochemical process of bio-hydrogen production is further subdivided into two different methods like (a) gasification and (b) pyrolysis. Mainly, gasification is a common method to produce bio-hydrogen gas. In this method, carbon-rich material is converted into bio-hydrogen thermochemically with/without catalyst at high temperature [50]. Energy is not released directly in the form of heat of combustion; rather it is converted in the form of bond energy as a lightweight fuel.

3.2.2 Bio-chemical method

The biochemical process of bio-hydrogen production is further subdivided into three different methods like (a) bio-photolysis, (b) dark fermentation, and (c) photo-fermentation. Nitrogenase, Fe-hydrogenase, and Ni-Fe-hydrogenase have commonly been used enzymes in the bio-chemical production method to convert biomass to bio-hydrogen [51, 52]. The bio-photolysis process is catalyzed by Fe-hydrogenase. This could be an indirect or direct process.

In the photo-fermentation process, a nitrogenase enzyme is used. By using water and nitrogen-deficient organic acids in the presence of light and nitrogenase enzyme, photosynthetic bacteria produce bio-hydrogen gas [53]. Dark fermentation is carried out by anaerobic bacteria by using carbohydrate-containing sources during the catabolism process of which carbohydrate-containing substrate bio-hydrogen gas is liberated [51].

4. Liquid bio-fuels

Liquid biofuel mainly includes biodiesel, which is of prime importance. Many techniques are in practice to produce the biodiesel.

4.1 Biodiesel

FAME (fatty_acid_methyl_ester) is a chemical name for biodiesel. It is a renewable bio-fuel and derived from the recycled greases, animal fats, and vegetable oils [54]. This fuel is just like petroleum-based diesel fuel and with very little to no reforms, it is safe to use in diesel engines also known as the compression ignition engine. Because of some important factors, biodiesel is superior over conventional diesel; for example, it does not contain sulfur residues, has lower life-cycle GHG emissions, and has lower particulate matter. Because of the high viscosity of vegetable oils, these can create problems; so we cannot use them directly. Vegetable oil has low volatility and its viscosity is 11–17× greater than the conventional diesel fuel; hence it forms deposits inside the fuel-injector of diesel engines because it does not burn completely [55]. The viscosity of vegetable oil can be reduced by following different methods. The most common production techniques are four in number (i) micro-emulsion, (ii) thermal-cracking also known as pyrolysis, (iii) dilution, and (iv) transesterification [56]. Transesterification is the preferable technique and it produces good-quality fuel. It is a chemical reaction between fat and oil by using alcohol mediums in the presence of a catalyst, which results in the formation of glycerol and esters [36].

4.2 Biodiesel: intensification techniques

As the use of fossil-fuel has influenced both human lives and the environment. Therefore, Biodiesel has revealed an optimistic effect to resolve environmental issues and helping to achieve energy requirements as a renewable energy resource [57].

Nonedible plants such as *Ceiba pentandra* (kapok), Jatropha, *Hevea brasiliensis* (rubber seed) should be preferred for the better quality of biodiesel production for social and economic values [58]. To find the easiest and economic way for biofuel production is the basic objective of the researchers. To enhance the ester product, the transesterification process should be fast and simple. Hydrodynamic cavitation, co-solvent, microwave heating, ultrasonic irradiation, the use of supercritical methanol, and in situ transesterification processes are novel applications to improve the biodiesel production [59]. According to literature, there are many studies about transesterification by microwave-assisted techniques using the nonedible plant-based oil.

4.2.1 Microwave-based method

A microwave-assisted technique reduces the processing time and also saves energy because the electromagnetic field is the main source of energy in this technique [60]. As to enhance the biofuel production and reaction rate in a very short period via a harmless and suitable method at optimum temperature, this microwave technique is of prime importance and is widely accepted as a chemical reaction tool [45]. In the transesterification method, catalysts like alkali convert the crude oil into fatty-acid-alkyl-esters (biodiesel) by using the monohydric-alcohol substrate. For biodiesel commercialization, the major issue is the operating cost.

To resolve this issue, a microwave-assisted technique as shown in **Figure 9** is the finest method to increase biodiesel yield and reduce reaction time; the transesterification method is an energy-saving and fast method for biodiesel production than the conventional method. This microwave method can make the separation process easier and also speed up the rate of reaction by providing specific heat during trans esterification reaction [46]. An increase of dipolar rotation phenomenon can reduce the activation energy when microwave electromagnetics interact with reaction components (triglycerides, alcohol) [61].

4.2.1.1 Factors influencing the reaction rate

The transesterification process is ruled by the amount and by type of alcohol. Methanol is one of the most preferable reactants that are used in both techniques of transesterification like microwave-based and conventional methods. Methanol

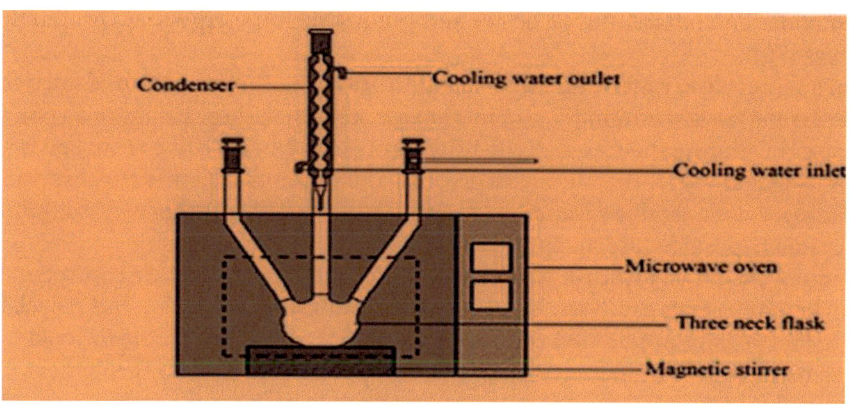

Figure 9.
Microwave setup for biodiesel synthesis.

is mostly used to achieve maximum production of biodiesel by using non-edible oils as those shown in **Table 4**. Methanol allows easier methyl ester formation and separation of glycerol simultaneously compared to other alcohols like ethanol [62]. So, this is because, in the catabolism of methane anoin ethoxide ($C_2H_5O^-$) have low reactivity compared to methoxide (methanolysis) [63]. The sensitivity of ethanol causes the formation of soap in the non-edible oil.

Vitally, methanol has more ability to absorb the microwave spectrum than ethanol under microwave irradiation. The electromagnetic energy storage in material measures the capacity of substances to insulate charges from each other in dielectric constant. The material becomes polarized easily in the electric field at high dielectric constant due to decrease in the dielectric constant through the growing straight chain in the R-OH. Methanol is more favorable for biodiesel production via microwave technique due to its smallest path between the hydroxyl and hydrocarbons group and a remarkable nucleophilic property.

Ethanol produced from renewable biomass is an appropriate product for the production of biodiesel; there are five types of alcohol in the transesterification process [64]. The obtained values show the ascending conversion order such as isopropyl-alcohol < 2,butyl-alcohol < iso-propenyl-alcohol < iso-butyl-alcohol < methanol. So, we concluded that methanol is one of the efficient absorbers compared to the iso-butyl and iso-propenyl alcohol in a microwave field. Though microwave technique is very efficient for a transesterification reaction, there is a barrier with the immiscibility of oils with alcohols, which direct the minor triglycerides to biodiesel conversion.

Hence, to achieve a better yield, increasing the reaction rate and catalysts solubility of alcohol is of prime importance [65]. These catalysts can catalyze the reaction both heterogeneously as well as homogeneously. Under microwave operating reaction requirement of a catalyst excluded about tenfold as compared to the conventional heating method in transesterification [62, 66]. Different catalysts such as alkaline (NaOH, KOH, CH_3ONa, CH_3OK) and acidic (H_2SO_4, HCl, H_3PO_4) are used in biodiesel production because of their highest performance [67].

Therefore, at the molecular scale of the microwave method, there is the inability for the suspended heterogeneous constituent part; so in this case, homogenous catalysts are more preferable. The catalysts ions help to increase the molecular interaction of solvents because of their strong electric field. So it is concluded that the rate of reaction of acidic catalysis is slower 4000 times than the basic catalysis reactions, therefore, acidic catalysis reactions are less common because of their acidic nature rather than the basic nature [68, 69]. Catalysts like sodium-methoxides and potassium are better and preferable for a large-scale production process [70].

In another case, water causes the saponification and the formation of fatty acids in the strong basic medium [71]. In methoxide solutions, these catalysts do not liberate H_2O during the transesterification process and behave like weak Lewis-bases. Negligible ester dissolution in glycerol and fewer yield losses are observable under methoxide catalysis transesterification reactions. In another way, sodium and potassium hydroxides are inexpensive catalysts.

Homogeneous catalyst's drawbacks of homogeneous catalysts are controlled by using heterogeneous catalysts. Heterogeneous catalysts are reusable and recyclable due to inexpensive production costs with very good performance rate to avoid undesired saponification reactions [72]. It is reported that heterogeneous acid base-catalyzed reaction can act as an intermediate in both transesterification and esterification processes and they provide easier separation and cleaner products at the same time. SRO, Al_2O_3/50% KOH, CaO, BaO, SiO_2/50% H_2SO_4 etc. are the best solid catalysts for the production of biodiesel [73–75].

DOI: http://dx.doi.org/10.5772/intechopen.92652

Microwave-assisted base-catalyzed transesterification reaction								
Feedstock	**Alcohol:Oil ratio**	**Catalyst (wt%)**	**Temp. (°C)**	**Time (min)**	**Power (W)**	**Conversion (%)**		**References**
Waste cooking oil	Methanol 6:1	CH_3ONa0.75%	–	3	750	97.9		[42]
	Methanol 6:1	NaOH 0.75%	–	3	750	96.2		[42]
	Methanol 6:1	NaOH 1%	64	5	600	93.36		[44]
	Methanol 6:1	KOH 2%	78	5	300	95		[57]
	Methanol 6:1	SrO 1.85%	Around 80°C	3	1000	93		[42]
Jatropha oil	Methanol 30:1	NaOH 4%	55	7	1200	86.3		[43]
	Methanol 7.5:1	KOH 1.5%	65	2	1200	97.4		[58]
Cottonseed oil	Methanol 6:1	KOH 1.5%	60	7	21%	92.4		[76]
Karanja oil	Methanol 33.4 wt%	KOH 1.33%	–	2.5	180	89.9	0.3	[59]
Palm oil	Ethanol 4:1	KOH 1.5%	–	5	70	97.4		[60]
	Methanol 20:1	K_2CO_3 3%	≈alcohol B.P	180	1000	85.63		[38]
	2-propanol 20:1	K_2CO_3 3%		180	1000	49.51		[38]
	Methanol 18:1	CaO 15%	–	4	900	96.7		[45]
Macauba oil	Ethanol 9:1	Novozyme 435 2.5%	30	15	–	45.2		[46]
	Ethanol 9:1	Lipozyme IM 5%	40	5	–	35.8		[46]

Table 4.
Catalysis of trans esterification reaction under microwave assisted technique.

In the conventional method, at the higher temperature crude oil viscosity decreases with an increase of reaction rate exhibiting positive influence and reaction time reduction in both microwave and conventional methods [77]. In microwave reaction system, rapid heat transfer is achievable by molecular interactions of ionic and dipolar compounds via electromagnetic field that dramatically accelerates the reaction rate and also produces heat energy by volumetric distribution compared to conventional method where heat is transferred via radiation. However, reaction yield increases with increasing irradiation power [78].

4.2.2 Ultrasonic sound waves-based method

While using solid and liquid mediums for the process of intensification ultrasonic method is preferable [79]. Mechanical energy is required for continuous mixing and initiation of reaction in transesterification process. Ultrasonic sound wave energy of higher frequencies can be used [80]. It speeds up the reaction and also enhances the biodiesel yield. Cavitation is a phenomenon that creates chemical and physical effects on the reaction [81]. Ultrasonic cycles generate and also increase the size of cavitation bubbles from ten to a hundred times bigger [82]. High temperature (5000°C) and pressure (500 atm) occur due to collusion of bubbles in less than a microsecond and lead to the vigorous mixing and rapid heating in the system between the two immiscible reactants shown in **Figure 10**. There are cycles of expansion and compression because of introducing ultrasonic waves to liquid. A positive pressure push liquid molecules closer in the compression cycle and negative pressure pull apart in expansion cycles [83]. Collision between the molecules generate radicals (H^+ and OH^-) promoting reaction rate between the reactants [84]. Hielscher Ultrasonic GmbH ultrasonic reactor (frequency: 18–20 kHz) for biodiesel production was introduced in 2000 [81, 85] and used low frequencies (28–40 kHz) for transesterification of vegetable oils. However, at 28 kHz there was a high yield production and reaction time was shortened considerably at 40 kHz [86].

Production of biodiesel from the oil of the Schleicher Triguga plant is influenced by ultrasonic irradiation [87]. The esterification reaction using H_2SO_4 catalyst reduced the acid value from 21.65 to 0.84 mg/g in the first step (reaction time: 20 min at 40°C). While a during a second step of transesterification reaction, a catalyst $Ba(OH)_2$ is used (80 min at 50°C) with the 96.8% conversion of

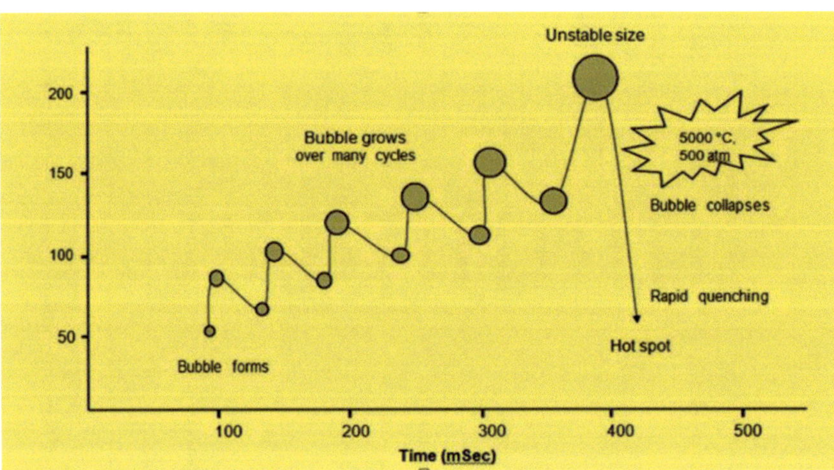

Figure 10.
Collapse and formation of bubbles inside the cavity.

Methods	Feedstock	Catalyst	Optimum reaction condition	Yield/conversion (Wt %)	References
MS	Waste cooking oil	KOH	Cat. 1 wt%, MeOH/oil 6:1, T 70°C, t 1 h	Y 98.2	[46]
MS	Used frying	NaOH	Cat. 1.1 wt%, MeOH/oil 7:1, T 70°C, t 0.33 h	Y 88.8	[47]
MS	Waste cooking oil	H_2SO_4	Cat. 4 wt%, MeOH/oil 20:1, T 95°C, t 20 h	C>90	[48]
MS	Rubber seed oil (42.5% FFA)	H_2SO_4	Cat. 10.74 wt%, MeOH/oil 10:1, T 65°C, t 60 min	FFA conversion 98.6%	[41]
MS	Rubber seed oil (40.14% FFA)	H_2SO_4	Cat. 7.5 wt%, MeOH/oil 23:1, T 50°C, t 30 min	FFA conversion 75%	[56]
UM	Rubber seed oil (40.14% FFA)	H_2SO_4	Cat. 7.5 wt%, MeOH/oil 23:1, T 50°C, t 30 min	FFA conversion 98%	[56]
UM	Soybean oil	KOH	Cat. 1.5–2.2 wt%, MeOH/oil 6:1, T 40°C, t 0.25 h	Y 99.4	[47]
UM	Waste cooking oil	KOH	Cat. 1 wt%, MeOH/oil 4:1, T 27–32°C, t 0.016 h	Y 99	[48]
UM	Oleic acid	H_2SO_4	Cat. 5 wt%, Ethanol/acid 3:1, T 60°C, t 2 h	C>90	[49]

Table 5.
Comparison between conventional stirring and ultrasonic irradiations.

triglyceride. So, in conventional methods, ultrasonic waves accelerate the transfer of mass in heterogeneous nature catalyzed reactions [88]. **Tables 5** and **6** show the comparison of the ultrasonic-assisted method as well as a method of conventional stirring for the transesterification reactions. In several studies, RSO (rubber seed oil) and non-edible oil is reported as acid pre-treated by using the ultrasonic method [89]. To harvest 98% of FFA conversion this reaction took only 30 min, while some researchers reports 60, and 90 min [90–93]. In mechanical stirring, observed FFA conversion was only 75%.

4.2.3 Bio-gasoline and green diesel: thermochemical upgraded processes

4.2.3.1 Hydro-deoxygenation catalytic process

For cracking of long-chain gas oils into small-chain petroleum product distillates, olefin hydrocarbons saturation, and removal of heteroatoms, catalytic hydroprocessing is the most preferable petrochemical method. Hydro-deoxygenation of vegetable oil occurs at a temperature of 300–450°C, with lowest space velocity and hydrogen pressure above 3 MPa. Vegetable oils are converted into saturated long-chain hydrocarbons with boiling points (180–360°C) [94]. HDO technology has been established by many commercial companies like HP-Innovations, UOP-Honeywell, Nest Oil-Corporation, Valero Energy-Corporation, and ConocoPhillips

Method	Feedstock	Catalyst	Optimum reaction condition	Yield/conversion (Wt %)	References
MS	Palm oil	KF/Ca-Al hydrotalcite	Cat. 4 wt%, MeOH/oil 12:1, T 65°C, t 5 h	Y 97.98	[51]
MS	Palm kernel	CaO	Cat. 6 wt%, Alcohol/oil 30:1, T 60°C, t 3 h	Y 98	[52]
MS	Palm oil	CaO/Al_2O_3	Cat. 6 wt%, MeOH/Oil 12:1, T 65°C, t 5 h	Y 98.64	[53]
MS	Soybean oil	15-KOH/CaO	Cat. 4 wt%, MeOH/oil 16:1, T 65°C, t 1 h	Y 97.1	[50]
MS	Soybean oil	Calcined sodium silicate	Cat. 3 wt%, MeOH/oil 7.5:1, T 60°C, t 1 h	Y>95	[57]
MS	Soybean oil	K_2CO_3/MgO	Cat. 1 wt%, MeOH/oil 6:1, T 70°C, t 2 h	Y 99	[58]
MS	Jatropha oil	Mg-Al hydrotalcite	Cat. 1 wt%, MeOH/oil 4:1, T 45°C, t 1.5 h	Y 95.2	[47]
MS	Waste cooking oil	K_3PO_4	Cat. 4 wt%, MeOH/oil 6:1, T 60°C, t 2 h	Y 97.30	[60]
MS	Waste cooking oil	MgO/TiO_2	Cat. 10 wt%, MeOH/oil 50:1, T 170°C, t 6 h	Y 91.6	[45]

Table 6.
Synthesis of biodiesel with heterogeneous base catalyst by mechanical stirring.

in green-diesel synthesis process. They successfully introduced the substantial potential of catalytic hydro-processing.

4.2.3.2 Factors affecting green-diesel production

Hydrogen consumption, properties, and rate of an exergonic reaction in liquid biofuels are greatly influenced by the feedstock [95]. It is stated that higher unsaturated and long carbon chain lipid containing feedstock cause undesirable aromatization, oligomerization, cyclization, and dimerization of unsaturated long-chain carbon compounds intermediates inhibits the catalytic activities [96]. In the hydro-processing technique, temperature range depends on the product of interest. Temperature is the most important factor to harvest the hydrocarbons in diesel range; so for this purpose, many researchers have done many experiments to optimize the temperature range for the production of hydrocarbon chains within the diesel range. For long-chain hydrocarbons, for thermal cracking process to yield lower carbon chains, best temperature is above 400°C.

5. Conclusion

With the increasing demand for energy, continually depleting fossil fuels and energy resources, and a substantial environmental risk because of the consumption of fossil fuels, there is a demand for alternative resource. These factors are not ignorable; keeping in mind this situation, many researchers have introduced biofuels as an alternative, sustainable, and renewable source of energy. These are producible at a very low cost by using organic raw material, feedstock, domestic waste, animal waste, plant waste, energy crops, seed oils, biological organisms, micro-algae, water algae, macro-algae, etc.; biofuels can be produced through different chemical and biochemical techniques and many advancements are being introduced day by day with new researches. Biofuels are the future of this universe due to their environment-friendly nature.

Author details

Tahira Shafique[1]* and Javeria Shafique[2]

1 Department of Biochemistry, University of Agriculture Faisalabad, Pakistan

2 Department of Biotechnology and Bioinformatics, Government College University, Faisalabad, Pakistan

*Address all correspondence to: rao_tahira@yahoo.com

IntechOpen

© 2020 The Author(s). Licensee IntechOpen. Distributed under the terms of the Creative Commons Attribution - NonCommercial 4.0 License (https://creativecommons.org/licenses/by-nc/4.0/), which permits use, distribution and reproduction for non-commercial purposes, provided the original is properly cited. BY-NC

References

[1] Guo M, Song W, Buhain J. Bioenergy and biofuels: History, status, and perspective. Renewable and Sustainable Energy Reviews. 2015;**42**:712-725

[2] Demirbas MF, Balat M, Balat H. Potential contribution of biomass to the sustainable energy development. Energy Conversion and Management. 2009;**50**(7):1746-1760

[3] Sadeghinezhad E, Kazi SN, Sadeghinejad F, Badarudin A, Mehrali M, Sadri R, et al. A comprehensive literature review of bio-fuel performance in internal combustion engine and relevant costs involvement. Renewable and Sustainable Energy Reviews. 2014;**30**:29-44

[4] Cruz CH, Souza GM, Cortez LA. Biofuels for transport. In: Chapter 11 in Future Energy. Second ed. London, UK: Elsevier Ltd; 2014. pp. 215-244

[5] Demain AL, Báez-Vásquez MA. Chapter 14: Biofuels of the present and the future. In: Suib SLBT-N and FD in C. Amsterdam, The Netherlands: Elsevier; 2013. pp. 325-370

[6] Zhang Y, Ghaly AE, Li B. Determination of the exergy of four wheat straws. American Journal of Biochemistry and Biotechnology. 2013;**9**(3):338-347

[7] Major P. Availability and physical properties of residues from major agricultural crops for energy conversion through thermochemical processes. American Journal of Agricultural and Biological Science. 2012;**7**(3):312-321

[8] Zhang Y, Yu X, Li B, Zhang H, Liang J, Li Y. Exergy characteristics of woody biomass. Energy Sources, Part A: Recovery, Utilization, and Environmental Effects. 2016;**38**(16):2438-2446

[9] Zhang Y, Wang Q, Li B, Li H, Zhao W. Is there a general relationship between the exergy and HHV for rice residues? Renewable Energy. 2018;**117**:37-45

[10] Suganya T, Varman M, Masjuki HH, Renganathan S. Macroalgae and microalgae as a potential source for commercial applications along with biofuels production: A biorefinery approach. Renewable and Sustainable Energy Reviews. 2016;**55**:909-941

[11] Zhang Y, Chen P, Liu S, Fan L, Zhou N, Min M, et al. Microwave-assisted pyrolysis of biomass for bio-oil production. In: Pyrolysis. Croatia: IntechOpen; 2017

[12] Chew TL, Bhatia S. Catalytic processes towards the production of biofuels in a palm oil and oil palm biomass-based biorefinery. Bioresource Technology. 2008;**99**(17):7911-7922

[13] Zhang Y, Zhao W, Li B, Li H. Understanding the sustainability of fuel from the viewpoint of exergy. European Journal of Sustainable Development Research. 2018;**2**(1):09

[14] Alaswad A, Dassisti M, Prescott T, Olabi AG. Technologies and developments of third generation biofuel production. Renewable and Sustainable Energy Reviews. 2015;**51**:1446-1460

[15] Pradhan A, Mbohwa C. Development of biofuels in South Africa: Challenges and opportunities. Renewable and Sustainable Energy Reviews. 2014;**39**:1089-1100

[16] Ghobadian B. Liquid biofuels potential and outlook in Iran. Renewable and Sustainable Energy Reviews. 2012;**16**(7):4379-4384

[17] Russo D, Dassisti M, Lawlor V, Olabi AG. State of the art of biofuels from pure plant oil. Renewable and Sustainable Energy Reviews. 2012;**16**(6):4056-4070

[18] Schulte LA, Ontl TA, Larsen GD. Biofuels and Biodiversity, Wildlife Habitat Restoration. Encyclopedia of Biodiversity. Second ed. 2013. pp. 540-551

[19] Pandey A, Larroche C, Gnansounou E, Khanal SK, Dussap CG, Ricke S, editors. Biomass, Biofuels, Biochemicals: Biofuels: Alternative Feedstocks and Conversion Processes for the Production of Liquid and Gaseous Biofuels. Elsevier GN-GEN Laboratories, Academic Press; 2019. p. 6

[20] Robertson BA, Doran PJ. Biofuels and Biodiversity: The Implications of Energy Sprawl. In: Encyclopedia of Biodiversity. Second ed. Waltham, MA: Academic Press; 2013. pp. 528-539

[21] Delshad AB, Raymond L, Sawicki V, Wegener DT. Public attitudes toward political and technological options for biofuels. Energy Policy. 2010;**38**(7):3414-3425

[22] Demirbas MF, Balat M, Balat H. Biowastes-to-biofuels. Energy Conversion and Management. 2011;**52**(4):1815-1828

[23] Saladini F, Patrizi N, Pulselli FM, Marchettini N, Bastianoni S. Guidelines for emergy evaluation of first, second and third generation biofuels. Renewable and Sustainable Energy Reviews. 2016;**66**:221-227

[24] Naik SN, Goud VV, Rout PK, Dalai AK. Production of first and second generation biofuels: A comprehensive review. Renewable and Sustainable Energy Reviews. 2010;**14**(2):578-597

[25] Akhlaghi S, Gedde UW, Hedenqvist MS, Braña MT, Bellander M. Deterioration of automotive rubbers in liquid biofuels: A review. Renewable and Sustainable Energy Reviews. 2015;**43**:1238-1248

[26] Ruan R, Zhang Y, Chen P, Liu S, Fan L, Zhou N, et al. Biofuels: Introduction. In: Biofuels: Alternative Feedstocks and Conversion Processes for the Production of Liquid and Gaseous Biofuels. Elsevier GN-GEN Laboratories, Academic Press; 2019. pp. 3-43

[27] Luna C, Luna D, Bautista FM, Estevez R, Calero J, Posadillo A, et al. Application of enzymatic extracts from a CALB standard strain as biocatalyst within the context of conventional biodiesel production optimization. Molecules. 2017;**22**(11):2025

[28] Joshi G, Pandey JK, Rana S, Rawat DS. Challenges and opportunities for the application of biofuel. Renewable and Sustainable Energy Reviews. 2017;**79**:850-866

[29] Gupta VK, Potumarthi R, O'Donovan A, Kubicek CP, Sharma GD, Tuohy MG. Bioenergy research: An overview on technological developments and bioresources. In: Bioenergy Research: Advances and Applications. Waltham, MA: Elsevier; 2014. pp. 23-47

[30] Franco CJ, Zapata S, Dyner I. Simulation for assessing the liberalization of biofuels. Renewable and Sustainable Energy Reviews. 2015;**41**:298-307

[31] Scaife MA, Merkx-Jacques A, Woodhall DL, Armenta RE. Algal biofuels in Canada: Status and potential. Renewable and Sustainable Energy Reviews. 2015;**44**:620-642

[32] Algayyim SJ, Wandel AP, Yusaf T, Hamawand I. Production

and application of ABE as a biofuel. Renewable and Sustainable Energy Reviews. 2018;**82**:1195-1214

[33] Noraini MY, Ong HC, Badrul MJ, Chong WT. A review on potential enzymatic reaction for biofuel production from algae. Renewable and Sustainable Energy Reviews. 2014;**39**:24-34

[34] Hao H, Liu Z, Zhao F, Ren J, Chang S, Rong K, et al. Biofuel for vehicle use in China: Current status, future potential and policy implications. Renewable and Sustainable Energy Reviews. 2018;**82**:645-653

[35] Li X, Luo X, Jin Y, Li J, Zhang H, Zhang A, et al. Heterogeneous sulfur-free hydrodeoxygenation catalysts for selectively upgrading the renewable bio-oils to second generation biofuels. Renewable and Sustainable Energy Reviews. 2018;**82**:3762-3797

[36] Lee KT, Lim S, Pang YL, Ong HC, Chong WT. Integration of reactive extraction with supercritical fluids for process intensification of biodiesel production: Prospects and recent advances. Progress in Energy and Combustion Science. 2014;**45**:54-78

[37] Fokaides PA, Christoforou E. Life cycle sustainability assessment of biofuels. In: Handbook of Biofuels Production. Duxford, UK: Woodhead Publishing; 2016. pp. 41-60

[38] Acheampong M, Ertem FC, Kappler B, Neubauer P. In pursuit of Sustainable Development Goal (SDG) number 7: Will biofuels be reliable? Renewable and Sustainable Energy Reviews. 2017;**75**:927-937

[39] Renzaho AM, Kamara JK, Toole M. Biofuel production and its impact on food security in low and middle income countries: Implications for the post-2015 sustainable development goals. Renewable and Sustainable Energy Reviews. 2017;**78**:503-516

[40] Taghizadeh-Alisaraei A, Hosseini SH, Ghobadian B, Motevali A. Biofuel production from citrus wastes: A feasibility study in Iran. Renewable and Sustainable Energy Reviews. 2017;**69**: 1100-1112

[41] Peter W. Biogas production: Current state and perspectives. Applied Microbiology and Biotechnology. 2010;**85**:849-860

[42] Montingelli ME, Tedesco S, Olabi AG. Biogas production from algal biomass: A review. Renewable and Sustainable Energy Reviews. 2015;**43**:961-972

[43] Christy PM, Gopinath LR, Divya D. A review on anaerobic decomposition and enhancement of biogas production through enzymes and microorganisms. Renewable and Sustainable Energy Reviews. 2014;**34**:167-173

[44] Wellinger A, Murphy JD, Baxter D, editors. The Biogas Handbook: Science, Production and Applications. Woodhead, EU Science Hub: Elsevier; 2013

[45] Amore KM, Leadbeater NE. Microwave-promoted esterification reactions: Optimization and scale-up. Macromolecular Rapid Communications. 2007;**28**(4):473-477

[46] Motasemi F, Ani FN. A review on microwave assisted production of biodiesel. Renewable and Sustainable Energy Reviews. 2012;**16**(7):4719-4733

[47] Balat H, Kırtay E. Hydrogen from biomass—Present scenario

and future prospects. International Journal of Hydrogen Energy. 2010;**35**(14):7416-7426

[48] Udomsirichakorn J, Salam PA. Review of hydrogen-enriched gas production from steam gasification of biomass: The prospect of CaO-based chemical looping gasification. Renewable and Sustainable Energy Reviews. 2014;**30**:565-579

[49] Saxena RC, Seal D, Kumar S, Goyal HB. Thermo-chemical routes for hydrogen rich gas from biomass: A review. Renewable and Sustainable Energy Reviews. 2008;**12**(7):1909-1927

[50] Moghadam RA, Yusup S, Azlina W, Nehzati S, Tavasoli A. Investigation on syngas production via biomass conversion through the integration of pyrolysis and air–steam gasification processes. Energy Conversion and Management. 2014;**87**:670-675

[51] Hallenbeck PC, Benemann JR. Biological hydrogen production; fundamentals and limiting processes. International Journal of Hydrogen Energy. 2002;**27**(11-12):1185-1193

[52] Peters JW, Schut GJ, Boyd ES, Mulder DW, Shepard EM, Broderick JB, et al. [FeFe]-and [NiFe]-hydrogenase diversity, mechanism, and maturation. Biochimica et Biophysica Acta (BBA)-Molecular Cell Research. 2015;**1853**(6):1350-1369

[53] Manish S, Banerjee R. Comparison of biohydrogen production processes. International Journal of Hydrogen Energy. 2008;**33**(1):279-286

[54] Banković-Ilić IB, Stamenković OS, Veljković VB. Biodiesel production from non-edible plant oils. Renewable and Sustainable Energy Reviews. 2012;**16**(6):3621-3647

[55] Demirbas A. Biodiesel production from vegetable oils via catalytic and non-catalytic supercritical methanol transesterification methods. Progress in Energy and Combustion Science. 2005;**31**(5-6):466-487

[56] Agarwal M, Chauhan G, Chaurasia SP, Singh K. Study of catalytic behavior of KOH as homogeneous and heterogeneous catalyst for biodiesel production. Journal of the Taiwan Institute of Chemical Engineers. 2012;**43**(1):89-94

[57] Bokhari A, Chuah LF, Yusup S, Klemeš JJ, Akbar MM, Kamil RN. Cleaner production of rubber seed oil methyl ester using a hydrodynamic cavitation: Optimisation and parametric study. Journal of Cleaner Production. 2016;**136**:31-41

[58] Bokhari A, Chuah LF, Yusup S, Ahmad J, Shamsuddin MR, Teng MK. Microwave-assisted methyl esters synthesis of Kapok (*Ceiba pentandra*) seed oil: Parametric and optimization study. Biofuel Research Journal. 2015;**2**(3):281-287

[59] Asif S, Ahmad M, Bokhari A, Chuah LF, Klemeš JJ, Akbar MM, et al. Methyl ester synthesis of *Pistacia khinjuk* seed oil by ultrasonic-assisted cavitation system. Industrial Crops and Products. 2017;**108**:336-347

[60] Thostenson ET, Chou TW. Microwave processing: Fundamentals and applications. Composites Part A: Applied Science and Manufacturing. 1999;**30**(9):1055-1071

[61] Perreux L, Loupy A. A tentative rationalization of microwave effects in organic synthesis according to the reaction medium, and mechanistic considerations. Tetrahedron. 2001;**57**(45):9199-9223

[62] Sajjadi B, Aziz AA, Ibrahim S. Investigation, modelling and reviewing the effective parameters in microwave-assisted transesterification. Renewable and Sustainable Energy Reviews. 2014;**37**:762-777

[63] Kanitkar A, Balasubramanian S, Lima M, Boldor D. A critical comparison of methyl and ethyl esters production from soybean and rice bran oil in the presence of microwaves. Bioresource Technology. 2011;**102**(17):7896-7902

[64] Muley PD, Boldor D. Investigation of microwave dielectric properties of biodiesel components. Bioresource Technology. 2013;**127**:165-174

[65] Mazo PC, Rios LA. Esterification and transesterification assisted by microwaves of crude palm oil: Homogeneous catalysis. Latin American Applied Research. 2010;**40**(4):337-342

[66] Borugadda VB, Goud VV. Biodiesel production from renewable feedstocks: Status and opportunities. Renewable and Sustainable Energy Reviews. 2012;**16**(7):4763-4784

[67] Lam MK, Lee KT, Mohamed AR. Homogeneous, heterogeneous and enzymatic catalysis for transesterification of high free fatty acid oil (waste cooking oil) to biodiesel: A review. Biotechnology Advances. 2010;**28**(4):500-518

[68] Fukuda H, Kondo A, Noda H. Biodiesel fuel production by transesterification of oils. Journal of Bioscience and Bioengineering. 2001;**92**(5):405-416

[69] Vyas AP, Verma JL, Subrahmanyam N. A review on FAME production processes. Fuel. 2010;**89**(1):1-9

[70] Atadashi IM, Aroua MK, Aziz AA, Sulaiman NM. The effects of catalysts in biodiesel production: A review. Journal of Industrial and Engineering Chemistry. 2013;**19**(1):14-26

[71] Singh A, He B, Thompson J, Van Gerpen J. Process optimization of biodiesel production using alkaline catalysts. Applied Engineering in Agriculture. 2006;**22**(4):597-600

[72] Gude VG, Patil P, Martinez-Guerra E, Deng S, Nirmalakhandan N. Microwave energy potential for biodiesel production. Sustainable Chemical Processes. 2013;**1**(1):5

[73] Encinar JM, González JF, Martínez G, Sánchez N, Pardal A. Soybean oil transesterification by the use of a microwave flow system. Fuel. 2012;**95**:386-393

[74] Azcan N, Danisman A. Alkali catalyzed transesterification of cottonseed oil by microwave irradiation. Fuel. 2007;**86**(17-18):2639-2644

[75] Zu Y, Zhang S, Fu Y, Liu W, Liu Z, Luo M, et al. Rapid microwave-assisted transesterification for the preparation of fatty acid methyl esters from the oil of yellow horn (*Xanthoceras sorbifolia* Bunge.). European Food Research and Technology. 2009;**229**(1):43-49

[76] Wang Y, Ou S, Liu P, Xue F, Tang S. Comparison of two different processes to synthesize biodiesel by waste cooking oil. Journal of Molecular Catalysis A: Chemical. 2006;**252**(1-2):107-112

[77] Leung DY, Guo Y. Transesterification of neat and used frying oil: Optimization for biodiesel production. Fuel Processing Technology. 2006;**87**(10):883-890

[78] Chen KS, Lin YC, Hsu KH, Wang HK. Improving biodiesel yields from waste cooking oil by using sodium methoxide and a microwave heating system. Energy. 2012;**38**(1):151-156

[79] Patil PD, Reddy H, Muppaneni T, Mannarswamy A, Schuab T, Holguin FO, et al. Power dissipation in microwave-enhanced in situ transesterification of algal biomass to biodiesel. Green Chemistry. 2012;**14**(3):809-818

[80] Mulet A, Carcel J, Benedito J, Rosselló C, Simal S. Ultrasonic mass transfer enhancement in food processing. Transport Phenomena in Food Processing. 2003;**18**:265-278

[81] Stavarache C, Vinatoru M, Nishimura R, Maeda Y. Fatty acids methyl esters from vegetable oil by means of ultrasonic energy. Ultrasonics sonochemistry. 2005;**12**(5):367-372

[82] Chowdhury P, Viraraghavan T. Sonochemical degradation of chlorinated organic compounds, phenolic compounds and organic dyes–a review. Science of the Total Environment. 2009;**407**(8):2474-2492

[83] Vajnhandl S, Le Marechal AM. Ultrasound in textile dyeing and the decolouration/mineralization of textile dyes. Dyes and Pigments. 2005;**65**(2):89-101

[84] Pang YL, Abdullah AZ, Bhatia S. Review on sonochemical methods in the presence of catalysts and chemical additives for treatment of organic pollutants in wastewater. Desalination. 2011;**277**(1-3):1-4

[85] Suslick KS. Sonochemistry. Science. 1990;**247**(4949):1439-1445

[86] Serrano-Ruiz JC, editor. Advanced Biofuels: Using Catalytic Routes for the Conversion of Biomass Platform Molecules. Apple Academic, CRC Press; 2015;**27**:120-125

[87] Sarve A, Varma MN, Sonawane SS. Optimization and kinetic studies on biodiesel production from kusum (*Schleichera triguga*) oil using response surface methodology. Journal of Oleo Science. 2015;**64**(9):987-997

[88] Sharma YC, Singh B. An ideal feedstock, kusum (*Schleichera triguga*) for preparation of biodiesel: Optimization of parameters. Fuel. 2010;**89**(7):1470-1474

[89] Trinh H, Yusup S, Uemura Y. Optimization and kinetic study of ultrasonic assisted esterification process from rubber seed oil. Bioresource Technology. 2018; **247**:51-57

[90] Ahmad J, Yusup S, Bokhari A, Kamil RN. Study of fuel properties of rubber seed oil based biodiesel. Energy Conversion and Management. 2014;**78**:266-275

[91] Chuah LF, Bokhari A, Yusup S, Klemeš JJ, Abdullah B, Akbar MM. Optimisation and kinetic studies of acid esterification of high free fatty acid rubber seed oil. Arabian Journal for Science and Engineering. 2016;**41**(7):2515-2526

[92] Mootabadi H, Salamatinia B, Bhatia S, Abdullah AZ. Ultrasonic-assisted biodiesel production process from palm oil using alkaline earth metal oxides as the heterogeneous catalysts. Fuel. 2010;**89**(8): 1818-1825

[93] Yadav AK, Khan ME, Pal A, Singh B. Ultrasonic-assisted optimization of biodiesel production from Karabi oil using heterogeneous catalyst. Biofuels. 2018;**9**(1):101-112

[94] Srifa A, Faungnawakij K, Itthibenchapong V, Assabumrungrat S. Roles of monometallic catalysts in hydrodeoxygenation of palm oil to green diesel. Chemical Engineering Journal. 2015;**278**:249-258

[95] Vouitsis I, Geivanidis S, Samaras Z. Liquid biofuels in Greece—Current status in production and research. Journal of Renewable and Sustainable Energy. 2014;**6**(2):022703

[96] Liu Y, Sotelo-Boyás R, Murata K, Minowa T, Sakanishi K. Hydrotreatment of vegetable oils to produce bio-hydrogenated diesel and liquefied petroleum gas fuel over catalysts containing sulfided Ni–Mo and solid acids. Energy & Fuels. 2011;**25**(10):4675-4685

Chapter 12

City-Scale Decarbonization Strategy with Integrated Hydroelectricity-Powered Energy Systems: An Analysis of the Possibilities in Guadalajara, Mexico

Dulce Esmeralda García Ruíz
and Jorge Alberto Navarro Serrano

Abstract

According to the UN, in the next 20 years, most of the world's population will live in urban areas. Cities consume a high amount of resources, between this water, for their sustenance, hence the greatest necessity of sustainable development plans. What viable options or strategies can we consider in Latin America such that it can resist the economic, political, and social changes that it is facing? Through prospective studies, in case of Guadalajara, it is possible to determinate how water can generate clean energy, and which are the other strategic areas to empower the city through decarbonization with an interoperative and smart loop system of co-benefits. This study can help in public policy decisions of medium-sized cities in Latin America.

Keywords: water, decarbonization, clean energy, metropolitan areas, co-benefits

1. Introduction

Today, the accelerated growth of cities and their demand for greater resources, and high levels of pollution and the lack of clear and reliable actions in planning systems took greater importance under principles established in worldwide objectives through sustainability, since its official publication in 1987. This created a bigger challenge to achieve a real global impact through the action at a local level.

The rise in the amount of CO_2 emissions in recent years is shown in the following chart of the World Bank (**Figure 1**).

According to the World Bank [1], Mexico is the country responsible for most of the CO_2 emissions in Latin America, most of which come from the consumption of liquid fuel.

Considering this, there is an urgent need for addressing this problem by taking action to diminish environmental pollution. Addressing this global issue through local action also generates awareness at local level as can be represented in the following illustration (see **Figure 2**):

 IntechOpen

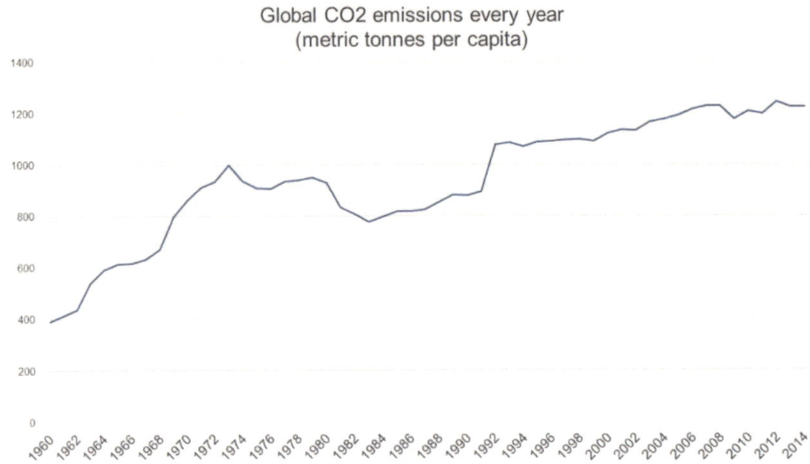

Figure 1.
Global CO_2 emissions by year (metric tons per capita). Source: Dulce García from World Bank, 2018.

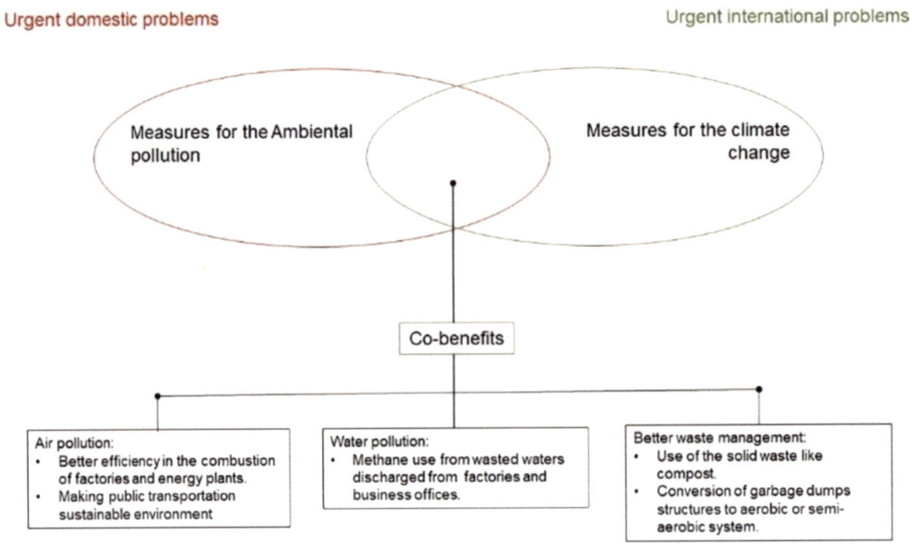

Figure 2.
Co-benefits, according of measures for the Ambiental pollution and the climate change. Source: From Approach is a new project, based in climate change concerns while also improving the local environment, 2009.

Meanwhile, co-benefits are a category of sustainability and were initially addressed by scientific literature in the 1990s as a response to public policies to anticipate climate change; in the literature, they have been understood as a measure of co-control that indicates that a single activity or policy can generate multiple benefits across several sectors or fields of study.

In many cases, the co-benefits are understood as the benefits of the climate, that is, based on the intentional decisions in which way it is possible to have benefits that have a positive impact on the environment.

Since 2007, the world has witnessed a new, historically radical reality: there are more people in the cities than in the countryside. The population that lives in the cities grows more and more. Therefore, there will be more energy consumption, which is increasingly responsible for the generation of global CO_2 emissions.

Between 1950 and 2005, the urban world population grew between 29 and 49%, and global carbon emissions jumped from 1630 to 7985 million tons ([2], p. 20).

Since the world's urban population will almost double by 2050, and most of that growth will take place in the developing countries ([3], p. 3), it is important to know and locate the necessary actions for addressing the problems faced by key cities and their metropolitan areas.

Guadalajara was chosen as the first case, since it is a city in the Latin American context with the second most important population, economic and territorial concentration in Mexico, which demands housing and services, thus amplifying the urban-environmental problem.

According to the City Prosperity Index CPI out by the UN in 2018, Guadalajara is at a basic level, this measure seeks the institutional alignment for the achievement of the objectives of Sustainable Development and the 2030 Agenda, this study shows that one of the main challenges facing the city in the category of environmental sustainability is within the category of air quality through the concentration of CO_2 with 46.07, as well as the treatment of wastewater with a percentage of 0.00 and the proportion of renewable energy generation with a percentage of 0.00, so it is essential to develop research with this orientation so that they can see the possibilities of decarbonizing Guadalajara through the generation of renewable energy through water as a measure of sustainable action [4].

It is also a city that attracts significant investment in industrial and technological matters; this central city had 1495 million inhabitants in 2010 (INEGI), together with the central nucleus of the metropolitan area that integrates a total of 4,434,878 inhabitants, this mind a contemporary urbanization with a regional influence of the phenomenon of metropolization [5].

Meanwhile, Jalisco is among the fourth largest energy consumer in the country [6–8] with some experts mentioning that the energy it produces is between 3 and 11% of the energy it consumes [7–9]. Only 3% of the energy consumed in the state is produced in Jalisco, with only 23% of the energy in Mexico coming from renewable sources, and 80% of the renewable energy coming from hydroelectric plants.

Likewise, one of the main objectives in Mexico is the development of renewable energy and the use of wind and solar farms; also, there is a lot of potential for the generation of electricity with the treatment of sewage water. Some experiments and prototypes have been developed as in the case of the one made by the Mexican Water Institute (IMTA) for the National Hydric Program (2014–2018) extracted from [10].

Specifically, in Jalisco, which is located within one of the four most populous states in the country, it is estimated that 83% of the electrical energy consumed is produced from fossil fuels and only 17% is generated from renewable sources. However, according to the data from the National Inventory of Renewable Energies (INERE), Jalisco has a high potential for renewable energy generation that has not yet been exploited. So, currently, in this situation, the energy transition model cut be promoting in projects to generate energy from various renewable sources within the state [11].

Therefore, in order to achieve this, not only clean energy but also energy efficiency in products is necessary; so, in Jalisco, a laboratory for the development of lighting technologies is being built to develop prototypes of luminaires with Internet of the things [12].

Additionally, another outstanding project is the one developed by the industrial digital company GE Power Systems, which is based on building the necessary infrastructure for the city through the creation of a power plant with HA turbines and GE digital services, which is estimated to be completed by the end of 2019; this project aims to generate 875 MV, enough to supply up to 2.8 million homes [7].

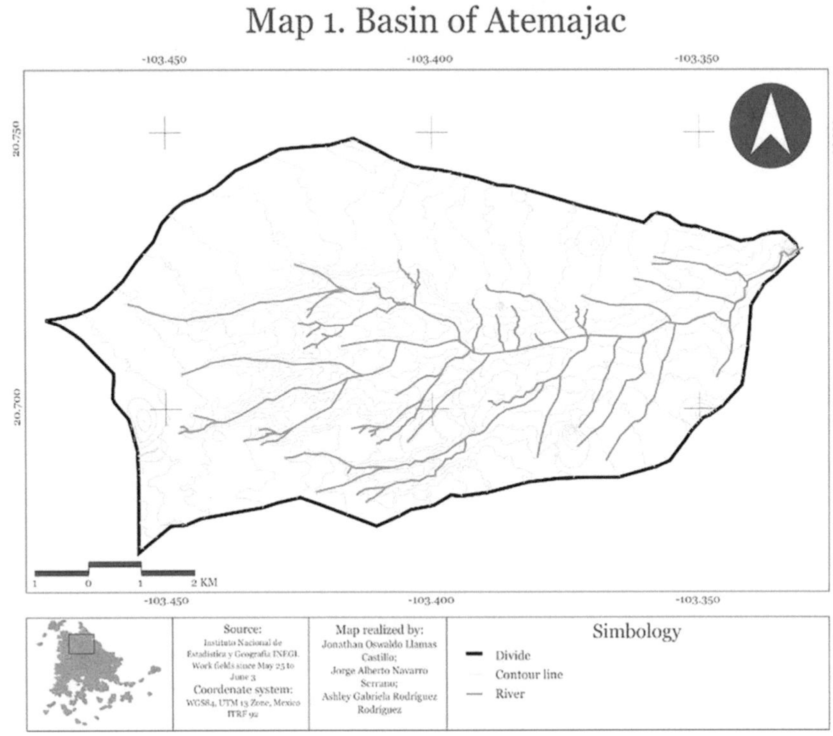

Figure 3.
Map of the basin of Atemajac and the topographical characteristics. Source: Jorge Navarro, Ashley Rodríguez and Jonathan Llamas.

It is necessary to have control actions that can help reduce or neutralize pollutant emissions, so that they meet the specific demand of the city and metropolitan areas, through their resources and services, which allows generating a balance between the environment, the reduction of social vulnerability, and a viable economic source of energy.

This research is focused on identifying through time retrospective diagnosis by means of cartographic analysis the potential of integration that exists in the water to generate clean energy as a measure of decarbonization in cities taking as a case study Guadalajara, Jalisco specifically the Atemajac river, with which a dynamic inter-operational model of the System could be established.

The study tried to verify if it is possible to integrate clean energy through water in the rivers by means of hydroelectric systems using decarbonization of the city and urban areas by attending to the specific case of the Atemajac river (see **Figure 3**).

2. Historical background

The strategical localization of Atemajac river between the capital of the kingdom of New Galicia, Guadalajara, and a little town like Zapopan, whose importance is, until the present, in a little statue, the Virgen de Zapopan, convert the river an important place in the cultural and economic situation. First, for the water like the most indispensable resource for all the activities, recreational and the "Romeria" a religious party between Zapopan and Guadalajara where the Virgen de Zapopan is taken in a way to the sanctuary. Thus, the Atemajac river is very important for social and cultural reasons, and for ecological reasons too. All of the above make this a river of much value for the region.

2.1 Precolonial and colonial period

Before the colonial period, in the Atemajac basin, there were towns formed by natives and under the government of the kingdom of Tonalá [13]; these are Zapopan, Zoquipan, and Atemaxac, all around the river and Mezquitan.

The existence of these towns will be altered by the arrival of the Spanish army to Atemajac Valley around 1530, when the social, political, and cultural situation would change. In this situation, Nuño de Guzmán found Guadalajara in 1539 in Nochistlan, today the state of Zacatecas for reach the coast of the gulf of California [14], but for the complicated situation in this place, the scarcity of water, the dry weather and other circumstances make the city move to Tonala in 1533, Tlacotan in 1541 and the Valley of Atemajac in 1542, in February 14 Guadalajara was founded in definitive way (**Figure 4**).

At the same time, also in the actual state of Zacatecas, the beginning of the war of El Mixton (named after the mountain in Apozol, Zacatecas) in 1540 [13] generated important changes in the territory. This war is one of the "most organized rebellions of Indian people in America" ([14], p. 23), and resulted in an important problem for the Spanish government, which will be fixed by Antonio de Mendoza, upon request by the government Cristobal de Oñate and the war cut over in 1541. After that, Mendoza give the order to exterminate the Indians who participate in the war, but, the intervention of Fray Antonio de Segovia and other persons like Francisco de Bobadilla, make a change in the decision of the viceroy and they cut live in the lonely towns, but under the Spanish laws and with the native and peninsular vision of the life, the social situation and the territory.

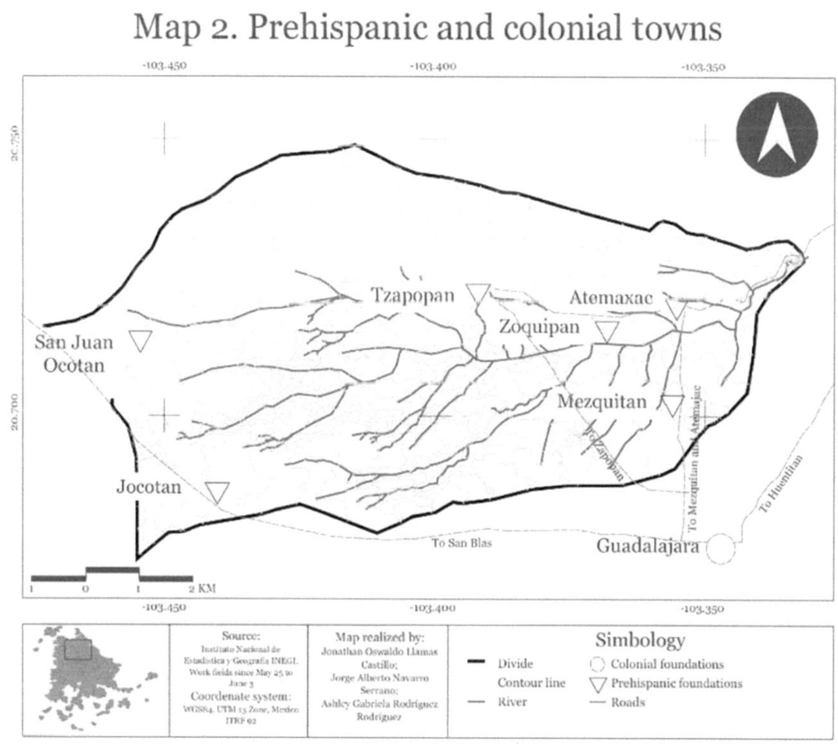

Figure 4.
Prehispanic towns like Tzapopan, Atemaxac, or Zoqiupan were present before the Spanish colonization, before the Spanish colonization, and after this process, Guadalajara was founded and the mixed of the Spanish and Indian vision make the territory. Source: Jorge Navarro, Ashley Rodríguez and Jonathan Llamas.

The water crisis in Guadalajara made the government look for a solution for this situation, and the Atemajac River seemed to be the perfect solution for this problem. Juan Rodriguez de Albuerne command to Pedro Buzeta to make a solution for the water crisis in Guadalajara [15, 16], the money is proved for a tax in the mezcal and other liquors [17], with this, Buzeta build an aqueduct under the land in tree branches, one of them, close of Colomos.

2.2 Century nineteenth, the factories and the energetical situation above the water

The independence of Mexico was declared in 1821, after which many presidents have been faced with complicated situations. In 1830, the minister of the interior and foreign affairs in the government of the president Anastasio Bustamante, Lucas Alaman, made a public polity about the industrialization in the country, for this, founded the Banco de Avio, an institution for a credits and support in the logistic and the machines [18]; the two principal objectives were: first, the financing of enterprises and industries for the creation of jobs and the stimulation of the interior market; and second, supporting the use of machines and new processes for production [19]. Meanwhile, one of the conditions for this support was the use of water as a source of energy, for its ease of use and because it was more cheap than coal.

But, in the state of Jalisco, the support for Banco de Avio never came; in this situation, all the businessmen formed groups for financing factories. Under this system, the first factory in the state of Jalisco was Jauja and Bellavista, in the city of Tepic, today the state of Nayarit.

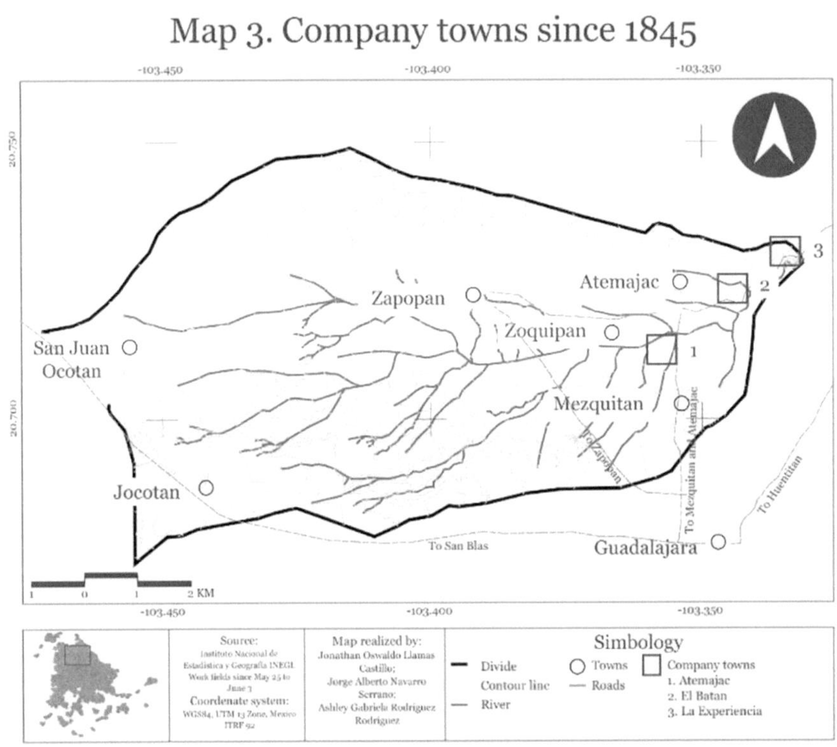

Figure 5.
The foundation of the company towns in the downside of the basin was possible for the existence of the river and a big market like Guadalajara and other cities, Atemajac, El Batán and La Experiencia take the energy of the water and give an industrial vocation to the river. Source: Jorge Navarro, Ashley Rodríguez and Jonathan Llamas.

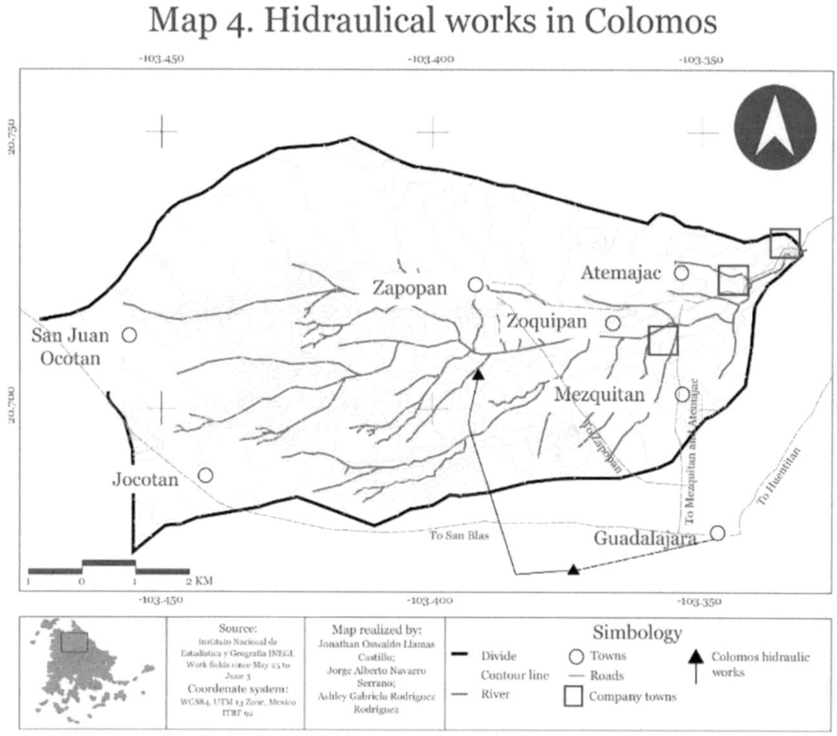

Figure 6.
At the end of the nineteenth century, the water of the Atemajac River was taken by the government for the city of Guadalajara from the place called Colomos. This generated the change in the energy fountain in the factories and the abandon of the Atemajac river for the population. Source: Jorge Navarro, Ashley Rodríguez and Jonathan Llamas.

Two important businessmen found three factories in Atemajac River, with water being the most important element to make energy and the Company Town being the model for urbanization around these factories. These factories are Atemajac and El Batan, the first dedicated to textile and the second to paper, both were founded by José Palomar in 1843 and 1844 [19].

In the case of Atemajac, Palomar built a dam in the place where the rivers Atemajac and Barranca Ancha or Culebras join, the "Zoquipan Dam" feeds a hydraulic wheel for the creation of energy for the factory. A similar situation happened in El Batán, with the own building in 1845 [9], besides a dam for a creation of the energy, Palomar build an aqueduct under the ground of stone for the transportation of the to make paper.

La Experiencia, a factory for textile, was founded in 1853 by Manuel Olasagarre [20]; this factory works with the energy produced by a falls where the San Juan de Dios River joins a little dam few meters up the river, with can be seen in the **Figure 2**. In the same place, in 1900, the CIJARA will build a hydroelectric to feed the three factories and a tram line from La Experiencia to Atemajac and Mexicaltzingo [21] (**Figure 5**). The construction of this plant is the result of a negotiation with the government and the "Compañia Industrial de Guadalajara, CIJARA" (Industrial Company of Guadalajara in English), if the water of Colomos will be introduced to Guadalajara (**Figure 6**), the government let to the company use the water of San Juan de Dios River for the generation of the electricity in the hydroelectrical plant. This situation makes a confrontation between the descendants of the CIJARA and the government [22], this situation still happens until today.

2.3 The grown of Guadalajara and the affectation in the river

The city of Guadalajara has seen an important growth since 1940, this was caused by the federal politics for the industrialization based on the substitution of importations, this consist in the fabrication of many product in the national factories for the local market, without a foreign product. Under this situation, many cities, besides Guadalajara like Monterrey or Mexico City, will grow.

The expansion of Guadalajara absorbs Zapopan and San Pedro Tlaquepaque. In the case of Zapopan, Atemajac River is the limit of both municipalities. First, the urbanization of the "Colonia Seattle" in the first decade of the twentieth century, second, the construction of the "Guadalajara Country Club" near of Zoquipan Dam and the construction for a road (Today Americas Avenue) between Zapopan and Guadalajara based in the old way [14], third, the construction of a new road, the actual Manuel Avila Camacho Avenue, and in last situation the construction of a park side of the Zoquipan Dam are an important fact for the urbanization of Guadalajara in the northwest side [14].

However, the affectation of Atemajac River cut be localized in this important constructions, first, the negotiation between Spanish acronym of Compañia Industrial de Guadalajara (CIJARA) and the government make an important controversy in Zapopan, the towns see in that time the river like a place of their property in the community perspective, this generate the first discharges of wastewaters, the construction of the Avila Camacho Park make possible the separation of wastewaters and rainwaters [23], also, both roads, the old and the new, the construction of Plaza Patria Mall in 1973 and Patria Avenue at the end of the same decade with the irregular neighborhoods in the downside of the river make the alteration of this.

Map 5. Guadalajara and the basin of Atemajac in the decade of 1940

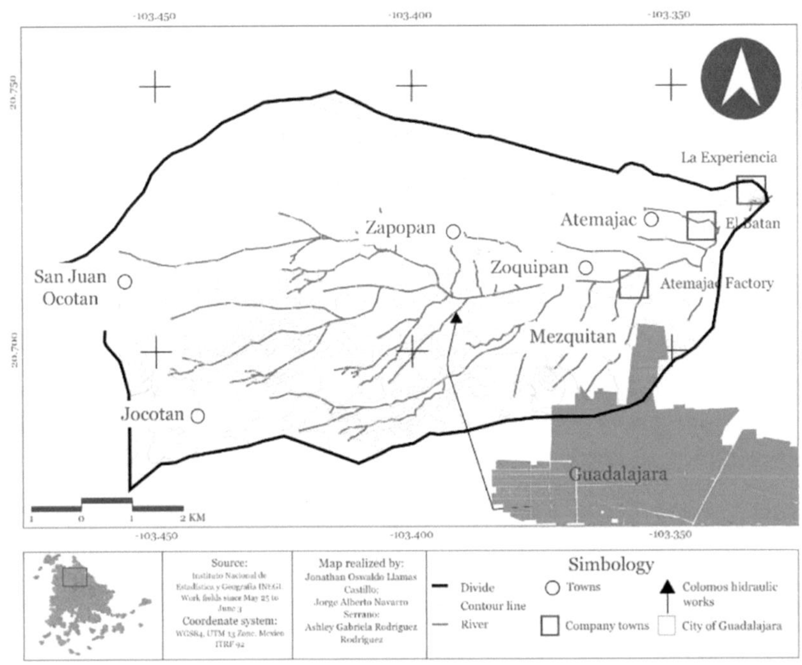

Figure 7.
From the decade of 1940 the city of Guadalajara begins to grow to the towns of Zapopan in the northwest and Tlaquepaque in the Southeast, this because of the third industrialization in Mexico, Mezquitan join to the urban area and the basin of Atemajac to. Source: Jorge Navarro, Ashley Rodríguez and Jonathan Llamas.

Today, the river is part of the Guadalajara Metropolitan Area, with 5,000,000 people; meanwhile, the Atemajac River is very rich in culture, history, and traditions, besides being an important antecedent in the creation of energy from water.

Figures 7–11 show the process of the growth of the city in the basin.

2.4 Dynamic modeling of the water-based clean energy system as a measure of decarbonization of the city

The methodology is based on the premise that dynamic systems change due mainly to the feedback cycles that occur between their elements, that is, their structure. The structure of the system is responsible for its behavior. This means that if you are looking for a profound change in a system, you must look at what changes are required in the structure, beyond addressing the symptoms of the problem. The structure of the system is represented in a causal system [24].

Likewise, the co-benefits in its categorical and conceptual badge are established within the scope of planning from public policies and up to its various scales of action as it is in programs; instruments; urban, architectural, environmental intervention projects, etc. And many areas or sectors such as the economic, social and environmental sector to achieve the reduction of public problems such as pollution [24].

Prospective studies in applied cities with effective public policy strategies as a measure of decarbonization of metropolitan areas.

Sustainability strategies based on co-benefits established in public policies as a decarbonization measure are duly located and articulated (under verifiable and quantifiable criteria) at various levels and sectors through prospective study systems in cities, which allow to easily detect the effects of pollution, as well as its action to meet the final objectives through water and clean energy generation.

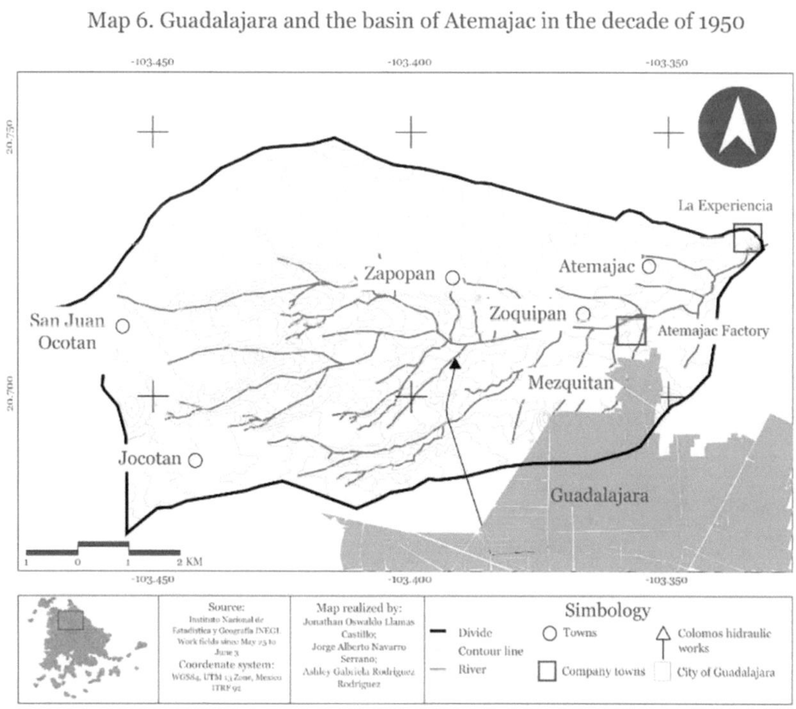

Figure 8.
In the decade of 1940, the growth of Guadalajara goes to the west, Atemajac Basin still is not part of the city. Source: Jorge Navarro, Ashley Rodríguez and Jonathan Llamas.

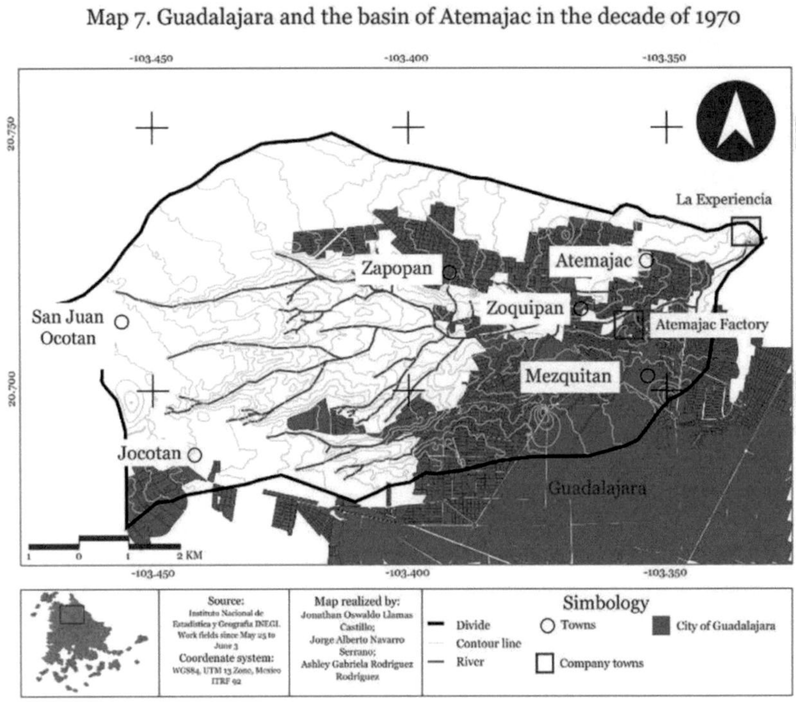

Figure 9.
Across the decade of 1970 the city is bigger, and many towns of the basin are included in the city area, Zapopan, and Atemajac converts in zones of the city and this will continue. Source: Jorge Navarro, Ashley Rodríguez and Jonathan Llamas.

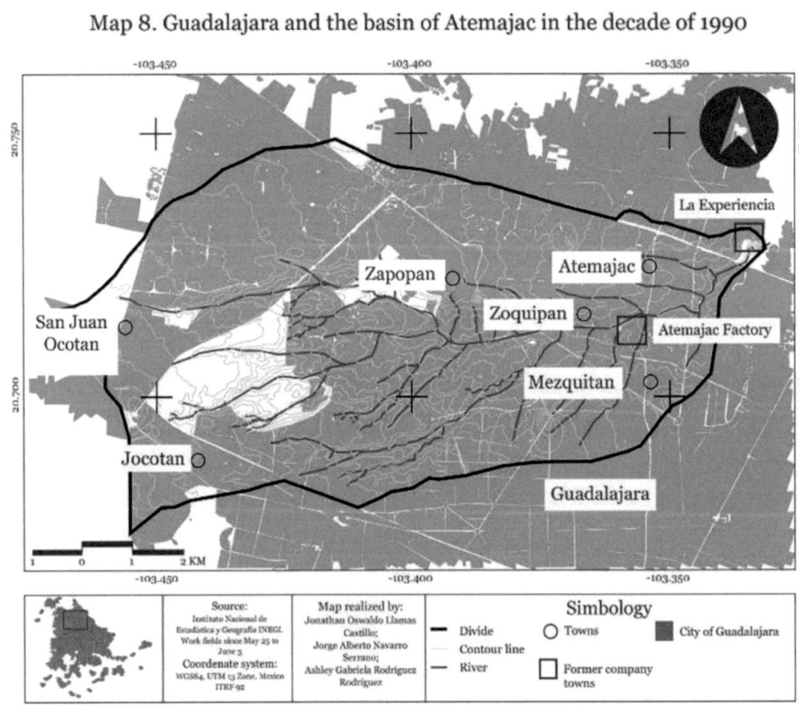

Figure 10.
Before of the twenty-first century, the basin is including in the Guadalajara Metropolitan Area, almost all the land is urbanized and the alteration in many rivers is present. Source: Jorge Navarro, Ashley Rodríguez and Jonathan Llamas.

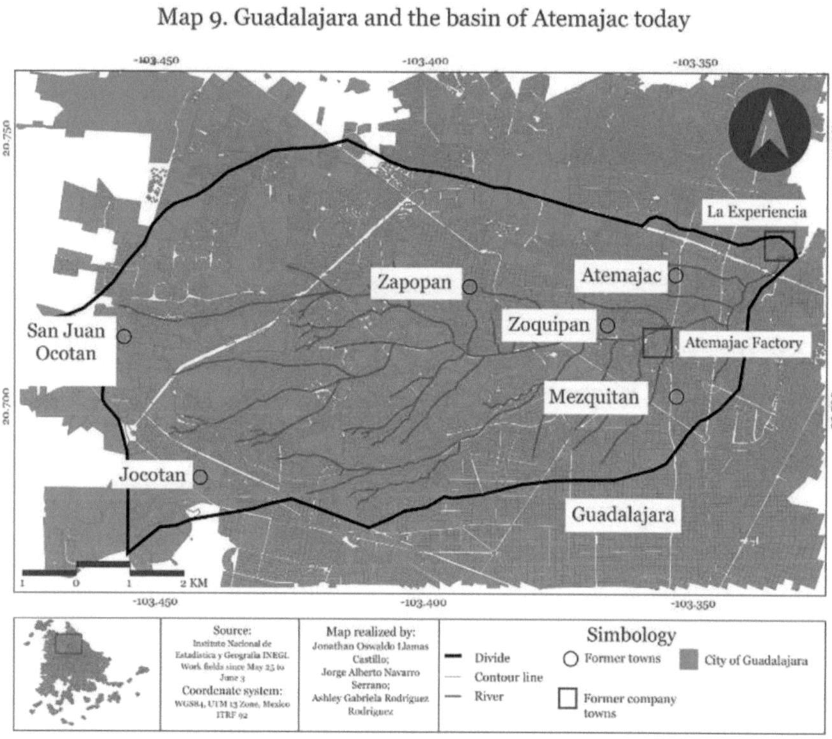

Figure 11.
In this day, the basin of Atemajac is all urbanized, inside there's old towns, new commercial environments and traditional neighborhoods, meanwhile, the city has made an alteration in the basin and the Atemajac river. Source: Jorge Navarro, Ashley Rodríguez and Jonathan Llamas.

There is a concern among government, society and companies at international, national, municipal and local levels about the relevance of having policies that foresee the effects of pollution in cities, which allows increasing the identification of variables through decarbonization strategies in an exponential way, not only in the planning systems, but also through the consumption habits of the population, until reaching the balance of the objectives searched in the System given in turn by the cycles of continuous feedback, however it is necessary constantly assessing population growth in metropolitan areas and their consumption habits which allows feedback to the system in an optimal and interoperable manner.

The decarbonization in the cities it's an idea than today is relevant because of the climate changing, the city is a complex system where the consumption of resources out of the territory is a tendency, one of them is the water, taken in the most cases of other basins; but, the possibility of take the water an energy in a local situation cut be possible [25].

The urban metabolism include the water like an important piece to moderate the problems with the climate change and the loss of the service; in this case, the city can use the water like a local source for the population and in the same scale, to make energy. It is important the information about the river and the basin, in this case the water flow in the channel, the dams for the protection, the rainwater catchment and the separation of the residual water to the clean water [26]. For a hydraulic energy obtentions, the cities cut recover the old infrastructure and build new to take this energy in a local perspective.

The possibility to obtain energy through water from the river of Atemajac cut be high, this because of the historical situation and the future in the channel, it minds, with a complete rescue of the space like a park, historical place and

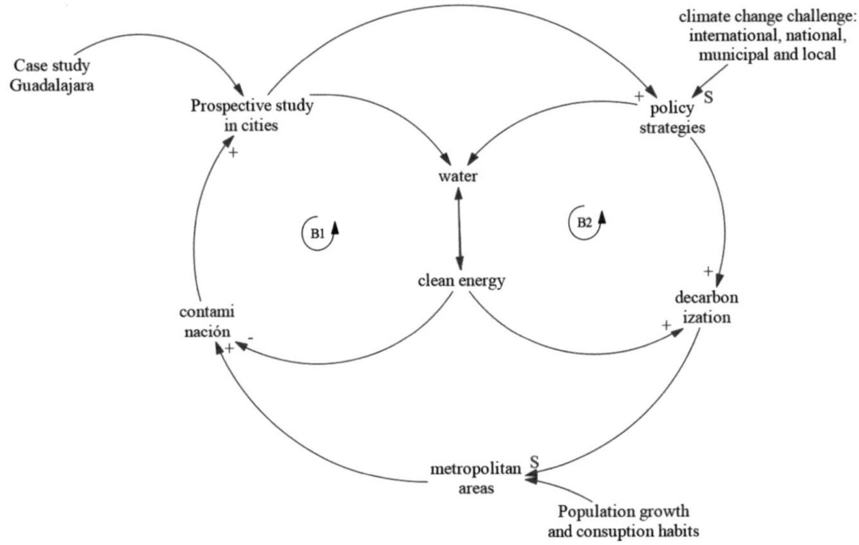

Figure 12.
General diagram about the co-benefits in the case of study by the policy strategies, decarbonization and other strategies in the metropolitan areas. Source: Author's elaborations by Vensim, 2019.

a source of water and energy. The clean water cut be good, but the wastewater also can serve [27], in the case of Guadalajara, both situations cut be possible, because of the existence of an infrastructure old and new for this, and the separation than the river have in both waters. Strengthened the actual infrastructure and recover and adapted the past will be important if Guadalajara want to have energy and come to the decarbonization from the water, special from Atemajac river. How can you conceptualize using **Figure 12**.

3. Results

Once all the variables that are immersed within the system have been linked, it is possible to model as follows by means of a simple causal diagram in **Figure 13**:

Based on the case study of Guadalajara, it is possible to say that through a co-benefits strategy in urban planning systems in cities, a co-control mechanism is generated which enables the generation of clean energy derived from the river water through hydroelectric systems, which in turn gives rise to self-sufficiency in production, involving a decarbonization mechanism of the city (**Figure 14**).

However, in the environment exogenous to the System, it can be found that urbanization and mobility of cities influence the natural course of rivers, which inhibits the energy capacity that can be derived from it through clean systems such as the hydroelectric plant. At the State and Federal levels, some strategies have been developed to be able to migrate to clean energies; it is important to integrate a joint strategy where the dynamic systems of which the city is a part are multi-operatized through strategic actions of co-control which feedback the Dynamic System until the desired balance is achieved.

Guadalajara has great potential for the reuse, reinjection, and disposal of rainwater, as well as giving it the correct treatment, which in turn enables the generation of energy through river water; however, by integrating the urbanization variable as this is one of the main causes of generation of CO_2 emissions due to the consumption and burning of liquid fuel, and being the runoff of the river inhibited by the urbanization of the city, which leads to this system becoming more complex

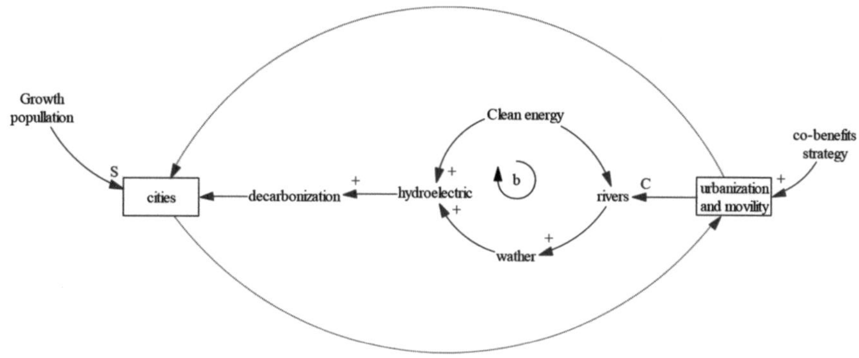

Figure 13.
General diagram about the energy and decarbonization in cities according to the water and his energy source. Source: Author's elaborations by Vensim, 2019.

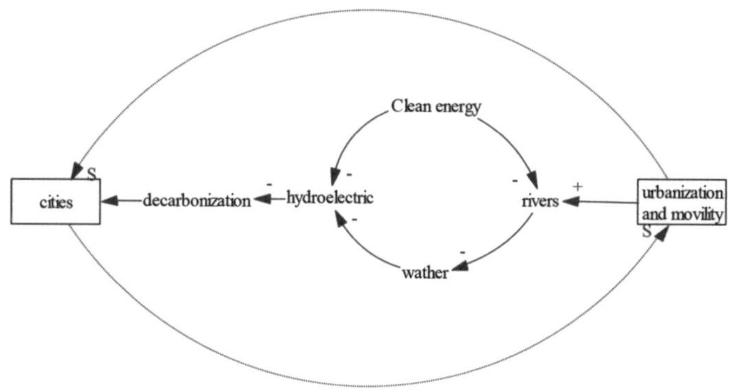

Figure 14.
General diagram about an anthropic system in the city. Source: Jorge Navarro personal files.

until generating an anthropic system without measure as can be seen in the following simple causal diagram.

Is type of growth is characterized by its two-phase transitional regime, one of them in exponential growth (urbanization) and the other in asymptotic decline (rivers). The positive feedback generated by exponential growth is narrowed by the negative feedback, which leads to the stabilization of growth. This is that every exponential process goes through a stabilizing process that limits growth. The above indicates that the exponential growth of this Sustained System does not exist in the real world (even with project that host certain initiatives in this regime as seen in the first section).

So a large increase in the positive variable in this case urbanization and mobility leads to a negative curve corresponding to the rivers, the negative curve does not appear spontaneously, on the contrary it is present at all times, but its effect depends on the influence of a variable in the positive curve (which in this case would be immersed in the planning systems and public policy that come from these prospective studies). When the positive curve begins to increase to all the variables involved in the cycle, the negative curve also increases until the domain changes and the negative curve is formed (as could be seen in the previous graph).

4. Attachment

In the next, there's some photography's about the work and walks in Atemajac river into the basin (**Figures 15–23**).

Figure 15.
Water birth in the high side of the basin of Atemajac. Source: Jorge Navarro personal files.

Figure 16.
Andares and Puerta de Hierro, the most rich and popular places in Guadalajara, localized in the high side of the basin of Atemajac. Source: Jorge Navarro personal files.

Figure 17.
In many places on the basin, there's an important and historical heritage with an important value, like this aqueduct from 1902 and projected by Agustin Pascal to transport water from Colomos to Guadalajara. Source: Jorge Navarro personal files.

Figure 18.
Construction of a dam clothes to Plaza Patria mall, this for contain the rains in the high side of the basin. Source: Jorge Navarro personal files.

Figure 19.
Zoquipan dam, in the past, the water conserved make the energy in Atemajac Factory, today is abandoned. Source: Jorge Navarro personal files.

Figure 20.
Atemajac river in the section of Patria Avenue and Enrique Diaz de León Avenue, clothes to Atemajac. Source: Jorge Navarro personal files.

Figure 21.
The river from Atemajac to Huentitán suffer a transformation in a polluted and dirty natural resource, the picture clothes to Alcade Avenue in the city of Guadalajara. Source: Jorge Navarro personal files.

Figure 22.
Colomitos was in the past a spa and resort, now, an urbanization reduces the territory of this water birth, know because of Pepe Guizar and his son "Guadalajara." Source: Jorge Navarro personal files.

Figure 23.
The river clothes Huentitán and Rancho Nuevo, the pollutions is present in the water and the air. Source: Jorge Navarro personal files.

5. Conclusions

It is important to understand, recognize, and relate action policies and their structure as an integral component of the city as a means to achieve sustainability, so that it is possible to locate and identify within them the planning systems. How can formulated the sustainable development of the context which the approach is proposed, in this occasion, is through the case of studies in Guadalajara.

Therefore, to verify that to achieve the imposed objectives it is necessary to have forecasting tactics that locate the indirect effects of the implemented strategies which implies that the objective of a policy or measure, generates a multi-operability and integration of the variables that compose it and the determination of the scope of the co-benefits and impacts produced for its correct implementation as a measure of decarbonization of the city through strategies that involve river water and energy.

The urban rivers and basins cut be the most important space to applicate the concept for water and energy to the decarbonization of the cities. For the source and make the energy by this recover old spaces (besides to be a recreative place, cut be a place to make energy) and strengthen the actual places. There is an important potential in the city of Guadalajara.

Notes/thanks/other declarations

Thanks to Ashley Gabriela Rodríguez and Jonathan Oswaldo Llamas Castillo for the contribution in cartography and other activities. Thanks to Diana Valeria Araiza Soto for her support in the correction and sharing information for preparing this chapter.

Thanks to Universidad Autónoma de Guadalajara for the support in this research.

Author details

Dulce Esmeralda García Ruíz* and Jorge Alberto Navarro Serrano*
Universidad Autónoma de Guadalajara, Guadalajara, Mexico

*Address all correspondence to: jorge.navarro@edu.uag.mx and dulce.garcia@edu.uag.mx

IntechOpen

© 2020 The Author(s). Licensee IntechOpen. Distributed under the terms of the Creative Commons Attribution - NonCommercial 4.0 License (https://creativecommons.org/licenses/by-nc/4.0/), which permits use, distribution and reproduction for non-commercial purposes, provided the original is properly cited. BY-NC

References

[1] World Bank. Emisiones de CO_2 (Toneladas métricas per cápita). Washington, DC: World Bank; 2018. Available from: https://datos.bancomundial.org/indicador/EN.ATM.CO2E.PC?end=2018&locations=MX&start=1960&type=points&view=chart [Consulted: July 25, 2019]

[2] de Souza CL, di Cesare Marques Awad J. Cidades sustentáveis, ciudades inteligentes. Desenvolvimiento sustentável num planeta urbano. Bookman: Sao Paulo, Brazil; 2012

[3] Ente SC, Navarro J, de Buen R, Cuevas P, de Buen O. Co-beneficios de un programa de vivienda energéticamente eficiente en México. Mexico City, Mexico: Comisión Nacional de Vivienda CONAVI, Environment Canada; 2013

[4] ONU Habitat, INFONAVIT. Índice básico de ciudades prosperas 2018. Mexico City, Mexico: ONU Habitat, Instituto del Fondo Nacional para la Vivienda de los Trabajadores INFONAVIT; 2018

[5] INEGI. México en cifras. Jalisco (14). Aguascalientes, Mexico: Instituto Nacional de Estadística y Geografía INEGI; 2010. Available from: https://www.inegi.org.mx/app/areasgeograficas/?ag=14 [Consulted: July 29, 2019]

[6] Jalisco Government. Área Metropolitana de Guadalajara. Guadalajara, Mexico: Jalisco Government; 2019. Available from: https://www.jalisco.gob.mx/es/jalisco/guadalajara [Consulted: July 29, 2019]

[7] General Electric. Guadalajara estrenará central eléctrica con turbinas HA y servicios digitales GE. Boston, United States: General Electric Reports Latinoamérica; 2017. Available from: https://gereportslatinoamerica.com/guadalajara-estrenar%C3%A1-central-el%C3%A9ctrica-con-turbinas-ha-y-servicios-digitales-de-ge-6305de64276b [Consulted: July 24, 2019]

[8] Patricia Romo. Jalisco apuesta por energías renovables. Mexico City, Mexico: El Economista Daily; 2017. Available from: https://www.eleconomista.com.mx/estados/Jalisco-apuesta-por-energias-renovables-20171101-0031.html [Consulted: July 24, 2019]

[9] El Informador Daily. Jalisco importa el 97% de su electricidad. Guadalajara, Mexico: El Informador Daily; 2014. Available from: https://www.informador.mx/Economia/Jalisco-importa-97-de-su-electricidad-20140310-0001.html [Consulted: July 24, 2019]

[10] IMTA. Generación de energía eléctrica a partir del tratamiento de aguas residuales por medio de bioceldas. Instituto Mexicano de Tecnología del Agua IMTA, Secretaría de Medio Ambiente y Recursos Naturales SEMARNAT; 2018. Available from: https://www.imta.gob.mx/generacion-de-energia-electrica-a-partir-del-tratamiento-de-aguas-residuales-por-medio-de-bioceldas [Consulted: July 24, 2019]

[11] Enrique Alfaro. Propuesta: 5. Transición energética. Enrique Alfaro; 2018. Available from: https://enriquealfaro.mx/territorio-y-suntentabilidad/sub-eje/5-transicion-energetica [Consulted: July 24, 2019]

[12] Patricia Romo. Jalisco desarrolla ocho proyectos energéticos. Mexico City, Mexico; 2018. Available from: https://www.eleconomista.com.mx/estados/Jalisco-desarrolla-ocho-proyectos-energeticos-20180831-0005.html [Consulted: July 24, 2019]

[13] Hayhoe AC. Guadalajara en el siglo XVI. Guadalajara City Hall: Guadalajara, Mexico; 1991

[14] María Muriá J, Olveda J, Rendón MA. Historia de Zapopan. Guadalajara, Mexico: El Colegio de Jalisco Press; 2004

[15] Cervantes MA. Agua para todos. La lucha de una ciudad por apagar su sed. Guadalajara, Mexico: Sistema Intermunicipal de los Servicios de Agua Potable y Alcantarillado, SIAPA; 1980

[16] Recio Á. El acueducto de Guadalajara y la obra de Fray Pedro Buzeta en España y Nueva España. Revista de Indias. 2016;**LXXVI**(268):717-749

[17] Cotilla ML. Historia de la introducción del agua desde su fundación hasta su fecha. Guadalajara, Mexico: Government of the State of Jalisco; 1960

[18] Durand J. La vida económica tapatía durante el siglo XIX. In: García LR, Robles MC, editors. Capítulos de la historia de la ciudad de Guadalajara. Guadalajara, Mexico: Guadalajara City Hall; 1992. pp. 41-58

[19] de la Torre F. Patrimonio industrial jalisciense del siglo XIX: Entre fábricas de textiles, de papel y de fierro. Guadalajara, Mexico: Government of the State of Jalisco; 2007

[20] Velarde FM. Historia de las fábricas textiles de Jalisco. Guadalajara, Mexico: Zapopan City Hall; 1992

[21] Gabayet L. Obreros somos. Diferenciación y formación de la clase obrera de Jalisco. Guadalajara, México: El Colegio de Jalisco Press, Centro de Investigaciones y Estudios Superiores de la Antropología Social CIESAS Occidente Press; 1988

[22] El Informador Daily. Interponen demanda contra el ejecutivo por predio de arroyo La Campana. Guadalajara, México: El Informador Daily; 2009. Available from: https://www.informador.mx/Jalisco/Interponen-demanda-contra-el-Ejecutivo-por-predio-del-Arroyo-de-la-Campana-20090202-0258.html [Consulted: June 26, 2019]

[23] El Informador Daily. El parque de Zoquipan. Guadalajara, México: El Informador Daily; 1965. Available from: http://hemeroteca.informador.com.mx/ [Consulted: June 26, 2019]

[24] Aracil J. Introducción a la dinámica de sistemas. Madrid, Spain: Alianza Editorial; 1983. Available from: http://tiesmexico.cals.cornell.edu/courses/shortcourse5/minisite/pdf/Literatura/Aracil%20Gordillo%20DS.pdf

[25] García AP. Energía y ciudad, un enfoque postambiental. Revista bibliográfica de Geografía y Ciencias Sociales. 2011;**XVI**(927). Available from: http://www.ub.edu/geocrit/b3w-927.htm

[26] Eicker U, editor. Urban Energy Systems for Low-Carbon Cities. London, United Kingdom: Academic Press; 2019

[27] Lazarova V, Choo K-H, Cornel P. Water-Energy Interactions in the Water Reuse. London, United Kingdom: International Water Association IWA; 2012